THE PRACTICE OF OCEAN RESCUE

A

Typical Towing Hook Assembly.

Acknowledgements to Cochrane & Sons, Ltd., Selby, Yorks.

(*a*) 'Liverpool' type Towing Hook with quick releasing gear. (*b*) Monarch Shock Absorber Unit.
(*c*) Bulkhead Hinge Block. (*d*) Bearing Ring. (*e*) Table Plate.

THE PRACTICE OF
OCEAN RESCUE

BY

R. E. SANDERS

GLASGOW

BROWN, SON & FERGUSON, Ltd., Nautical Publishers

52 Darnley Street

A*

Published **1968**
Revised   **1977**

ISBN 0 85174 294 7

© 1977 Brown, Son & Ferguson, Ltd., Glasgow, G41 2SG

Printed and Made in Great Britain

FOREWORD

The first Section in the book was included at the request of the Late James Ryan, Esq., G.M. Having spent a lifetime in tugs of all sizes and types, Mr. Ryan was a veritable mine of information concerning them and would, under different circumstances, have produced a much more comprehensive account of the tug type than the very brief account that is submitted here.

The second section aspires to satisfy what appears to be a fairly general desire for an outline upon the construction and interior economy of the ocean-going tug type.

No apologies are offered for a fairly extensive Section III, seeing that it is generally held among seamen that there is ordinarily not nearly enough information readily available concerning marine cordage.

In the compilation of Section III, the kind help of both Cdr. L. Hollis, M.B.E., R.N. (Ret'd) and David Rhys Davis, Esq., are gladly acknowledged in the search for an answer to the problem of sag values. It is not claimed that the solution provided is the best one or the only one. It is, however, one that gives a reasonable approximation easily and quickly.

Section IV aims to remedy some of the omissions and deficiencies of a paper originally written at the request of Commander E. G. Martin, O.B.E., V.R.D., R.N.V.R. (C.R.T.B. Campbeltown, 1939/46) and which was subsequently modified for inclusion elsewhere. The Chapter on Yaw also represents a re-write of an earlier paper written at the request of Captain G. S. Holden, R.N. (C.C.R.T. Admiralty, 1943/46). This section also suggests practical usages in the employment of ocean tugs in the close quarters role which some will instantly recognise as being those advocated by Lt.-Cdr. T. Bond, R.N.R.

v

THE PRACTICE OF OCEAN RESCUE

Sections V and VI result from the influence of Captain Louis Colmans who has extensive experience of both ocean towage and marine salvage. Captain Colmans has, for many years, been a strong advocate of team effect in salvage dispositions.

Finally, it must be remarked that this work does no more than to reflect the practical instruction and basic teaching which was so willingly provided, many years ago by a most notable gathering of seamen which, among others, included:—

> Captain G. H. Spence, O.B.E.
> Captain Teum Vet, M.B.E., D.S.C.
> Lt.-Cdr. Timothy Bond, R.N.R.
> Captain Jan Kalkman, M.B.E., D.S.C.
> Captain Peter Kent.
> Captain (Uncle) Ben Veltevreden.
> Lt.-Cdr. O. Jones, O.B.E., R.N.R.

ACKNOWLEDGEMENTS

The Navy Department of the Ministry of Defence for their help in several aspects of the preparation of this book.

The British Standards Institution for permission to reprint data from their publications.

Messrs. British Ropes, Ltd., of Doncaster, in the persons of M. B. Christie, G. M. McLean and A. B. Watts for their most generous and expert advice.

Messrs. Clarke Chapman & Co., Ltd., of Gateshead-on-Tyne, for the fullest possible co-operation in the Chapters devoted to the Towing Winch.

Messrs. Cochrane & Sons of Selby, for Towing Hook Assembly Detail.

S. S. Cornish, Esq., Librarian, Essex County Council, for a very great deal of assistance with this and other projects.

Messrs. Merryweather & Sons, Ltd., of Greenwich, for their very willing help with salvage pump and fire-fighting information.

Messrs. Siebe Gorman & Co., Ltd., of Chessington, in the person of Miss Audrey Scott, for valuable aid with Section V.

Lloyd's of London for permission to reprint *Lloyd's Open Form*.

The Monarch Towing Hook Co., Ltd., of London, for their help with towing hooks.

Messrs. W. W. Greener, Ltd., of Birmingham, in the person of Leyton Greener, Esq., for a deal of information upon the projection of lines.

CONTENTS

SECTION I

PAGE

SECTION II

SECTION III

Illustrations by the Author

SECTION I.

CHAPTER I.

The Evolution of the Ocean-Going Tug Type.

ONE of the first successful applications of the principles of auto-propulsion to the commercial marine requirement lay in the introduction, towards the end of the first half of the nineteenth century, of the pioneer steam tugs. The notion of applying steam power in this fashion had occurred, almost simultaneously, to shipping interests in most of the major Northern European Ports and, during the years 1835 to 1845, there appeared, on the waters of the principal seaports, a variety of small steam propelled craft which were designed to assist sailing vessels into and out of port and in their movements within the confines of the rivers, docks and harbours. This innovation was an unqualified success and much of the time which had hitherto been spent in laboriously working sailing vessels through port approach channels, not to mention the complicated warping processes required to shift vessels in port, was saved for more economic use at sea.

These first tugs were comparatively minute vessels, rarely exceeding sixty feet in length and constructed, in the main, from wood. They were propelled by a pair of paddle wheels fitted with fixed wooden floats rotated by a single cylindered steam engine deriving power from a vertical flue boiler of exactly the same pattern as was coming into general use in shore power installations. Condensers, when they were deemed necessary, were of such extravagant and primitive design that the endurance of most of these early tug types was measured in hours rather than days. Reserve feed water was carried in a tank adjacent to the boiler, and was transferred by hand pump. Bunkers, rarely exceeding a total of three or four tons, were dumped within shovel-reach of the furnace door whilst reserves, when considered necessary, were stowed as convenience dictated, usually on deck.

The towing effort of these pioneer craft was transmitted to the tow either through a large wrought iron hook secured by bolts, or through rivetting, to the after side of the boiler casing, or through a post of appropriate dimensions stepped on to the keelson and partnered at the deck in the same fashion as had been found effective for the fitting of sailing vessel's masts.

FIG. 1

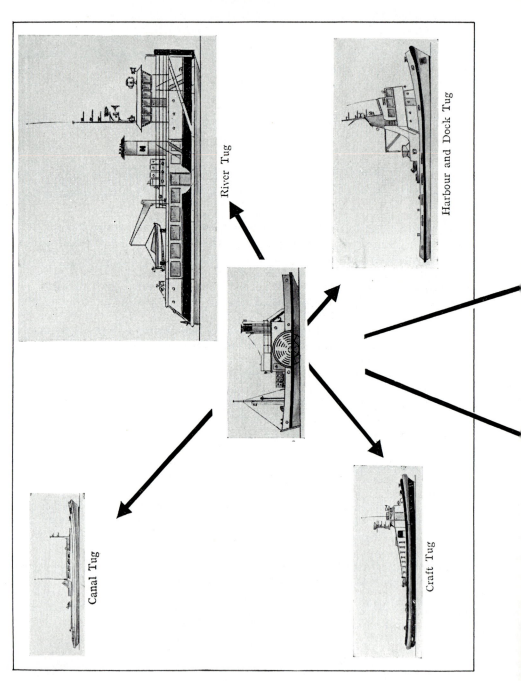

Canal Tug

River Tug

Harbour and Dock Tug

Craft Tug

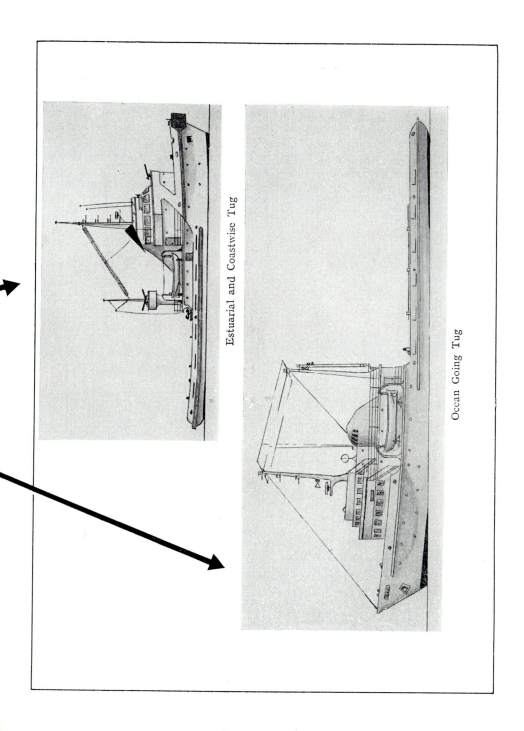

Estuarial and Coastwise Tug

Ocean Going Tug

The performance of these original types was studied most attentively by the representatives of a number of separate maritime interests, each of whom saw an application of this power towing principle to his own particular requirement, so that in due course there was evolved, from this basic tug type, the variety of specialised towing craft which are in general use today namely, the River and Canal Tugs, the Craft Tugs, The Harbour and Dock Tugs, The Estuarial and Coastwise Towing Types and lastly the type with which this book concerns itself, the Ocean-Going Tug Type. (Figure No. 1).

CHAPTER II.

The Ocean-Going Tug

During the second half of the nineteenth century a vast international fleet of deep-sea sailing vessels, already in intense competition with one another in the cargo and passenger trades between European Ports, the Americas, and the rapidly developing new countries and colonies overseas also the Orient, found added competition from a steadily increasing fleet of steam propelled vessels. The disparity between the respective open-ocean performances of the sailing vessel and the steamer was not great, particularly in the very early days of steam; but in the narrow seas and coastal waters, where manoeuvring space was restricted, the onset of adverse winds imposed a most crippling disadvantage upon the sailing vessel. The steamer's advantages were even more marked where a port's approaches involved the negotiation of a channelled estuary and a greater or lesser extent of river navigation. In such case a sailing vessel, under adverse circumstances, could spend as much time beating about in the approaches to a port as she had spent on the ocean passage.

The Sailing Ship Owner, in endeavouring to remain in competition and desirous of taking all possible profit from the type of vessel in which he was heavily committed financially before it succumbed to the obvious potential inroads of powered propulsion, decided to extend the range of towage from the immediate vicinity of the port to a point as close as practice would permit to the end of the open-seas voyaging. Such towage assistance, later to be generally known as 'Taking Steam,' obviously indicated some considerable modification to the basic tug type.

Quite clearly a ship of more substantial proportions was required in order to provide a hull of sufficient seaworthiness for the open-seas character of the work envisaged. A bigger ship would be needed for the quantity of bunkers, feed water and stores involved. It followed too that an urgent application was needed in order to improve upon the economy and efficiency of existing power-plants; there appeared to be a whole field of separate development in the matter of condenser design alone, before the projected type of operation could be seriously tackled. The essential qualities required of the new type were however rapidly recognised and the work was put in hand to produce craft having the following characteristics:—

1. Adequate size and seaworthiness.
2. Appropriate Endurance.
3. Sufficient Towing Power.
4. Economical and efficient machinery Installations.

In this, nothing has changed. The tugs designed and built one century and a quarter later still need these same basic qualities.

In the year 1855 the first real sea-going tugs went into service. These pioneers were barely effective, they required the assistance of sail in order to eke out their limited reserves of bunkers and their towing power was such that sailing ship Masters viewed their efforts at assistance with a critical scepticism. Not only had specialised tugs to be developed but a whole new seamanship technique had to be worked out to make an effective use of the new craft coming into service. However, with the experience of actual operations the tugs were modified and improved and the proficiency of the crews, particularly the Masters, was enhanced with the completion of each successful operation. During the ten years which then ensued, the term 'Seeking' became the designation for a whole sub-industry based upon the declining years of sail propulsion.

From the 1860's right up until the outbreak of World War I, tug operators based at the principal British and North European Ports despatched tugs down to the Western Approaches to seek out and assist the sailing vessels into their respective ports. Under the spur of healthy competition, operators steadily improved the tugs, whilst the crews became adept in the handling of their craft whilst towing and in handling the large type of cordage associated with the work. These were the formative years of sea-going tug development; the essential qualities having been correctly assessed from the outset of events, operators were able to effect steady improvement with every new unit brought into service. Wooden construction gave ground to iron and steel in succession. Hulls were modified, enlarged and strengthened as and where experience dictated. Paddle wheels were strengthened for sea service and were fitted with feathering devices to enhance float efficiency, sponsons and paddle box assemblies, which were proving to be sources of peculiar vulnerability, were more robustly constructed and were ultimately completely integrated into the principal hull components. With the experience accruing from a rapidly expanding engineering industry ashore, boiler and condenser installations were improved, whilst it became the custom to fit sea-going tugs with two, and upon occasion even three or four, separate boilers to provide the dual flexibility required for seeking and towing. The original single cylinder engine was replaced by multi-cylinder units which ultimately gave place to a twin-engined installation which gave the improved manoeuvring qualities which experience was indicating as requisite.

The deep-sea tug operator was however tardy in his acceptance of the marine screw propeller. Long after the screw propeller had received universal approval from the ordinary run of marine interests, the tugs clung to the paddle. This was partly because the operator had invested heavily in paddle development, and partly because of the innate conservatism of the sea-going tugmasters, who feared a loss in manoeuvrability with the adoption of the screw. It was also a fact that in their initial experimenting with screw propulsion for towing, the operators had failed to appreciate the need for some increase in the proportionate designed draught of screw tugs. Defective design, in this particular aspect, produced a few very expensive failures, when ineffective screw immersion in heavy weather ruined a few reputations and emptied a like number of purses. It was therefore not until the last decade of the nineteenth century that such inhibitions were overcome and screw tugs began to supersede paddlers in all towage functions.

Tug owners in all of the North European Ports had appreciated that the 'Seeking' function of the sea-going tug type was only an interim feature of marine employment from the very outset. It had always been recognised, by all who had taken part in the business, that the employment would only obtain until steam finally took over completely from sail, so that when this happened, in the first decade of the twentieth century, most tug operators had laid out an effective sheet anchor in the shape of a fleet of river, dock or harbour tugs. Indeed, the growing size of steamers had helped in the matter seeing that their increasing bulk was demanding inshore tugs of a size and power which had only recently been visualised solely in relation to the seeking function. So, as steam completed its ascendancy on the high seas, the tugowner reduced his sea-going fleet and built up on the inshore application.

It was however the truth that whilst, during this period of tug development, the seeking function had become the principal occupation of the sea-going tug, this type of vessel had carried out a number of highly successful rescue and salvage operations, some of which are regarded, to this day, as having been of quite epic character, this despite the fact that such tugs were, for the hazards of the salvage and rescue function, still comparatively frail and quite definitely under-powered. Such operations were not, at that time, regarded as being of any particular development potential, largely because salvage values were still comparatively low and not nearly sufficiently attractive to warrant any substantial deviation from the 'bread and butter' business of providing steam assistance to sail. With the end of the seeking employment in sight then, it seemed that the employment, in any quantity of the sea-going tug type, was drawing to a close. So, in fact, it would have been had it not been for one significant factor . . . 'Steam was not all that it was cracked up to be . . .'

The early steamers were, by any standards, quite under-powered for the deadweight demands imposed upon them. Because of their stage of development in a comparatively new era they were greatly prone to mechanical defect and failure and to occasional complete disablement. Shipowners who had, for centuries, become accustomed to the mild expenditure involved in an annual careenage, slipping, or more latterly, the drydocking of their ships, where the repair bill mostly resulted from weather damage and where maintenance work could be carried out by officers and crew, did not take too kindly to the expense involved in the maintenance and repair of a power installation. Ships were not therefore too effectively maintained and they were not very good to begin with; there was then, still work for the sea-going tug in the reclamation of the crippled. This provided a compensatory harvest for those operators who had, for one reason or another, remained in the deep-sea towing business. Because this reward was shared by a substantially reduced fleet, the individual return was sufficient to merit the modification of the sea-going tug type to meet the demands of the new function.

As steamers grew in size and became progressively more complex, their insured value rose until a total salvage award became a prize of great value, furthermore the obvious tangibility of the prize removed it sufficiently far from a gamble that it merited the acquisition and organisation of an effective fleet of tugs to guarantee the winning of the prize. The competition for such prizes induced operators to design and build the fore-runners of the very specialised ocean-going vessels in use today. Tugs were put in hand which were capable of proceeding beyond the narrow seas out into the open ocean to make progress against the weight of winter wind and sea in order to locate and tow home vessels crippled by mechanical breakdown, weather or other hazard. The more imaginative of the ocean-going tug operators were among the very first shipowners to have their vessels equipped with the Wireless Telegraphy installations which, in the first few years of the century, had already proved to be quite invaluable to the sea service.

It was, in all manner of ways, unfortunate for Great Britain that so many of her tug operators had elected to retire from the sea-towing sphere of operations with the decline of the seeking function. In the years, so soon to pass, which brought the First World War upon her, Britain failed to build and man the Ocean-Going tugs which could have saved some hundreds of ships and thousands of sailors' lives, that were literally cast away for want of so little foresight at the right level.

When, so soon after the outbreak of war in 1914, the Imperial German Navy declared a policy of unrestricted U-boat warfare, the sinkings of British and Allied Merchant Ships rose to such proportions that the very survival of the nation hung upon this single circumstance alone. Ships were being sunk, quite literally, faster than we could replace them. Even worse, many ships were hit but still floated, but could not be brought home to be repaired for further service. Some such vessels actually floated around for weeks before they were finally despatched by a U-boat desiring the credit for a kill. Others were deliberately sunk by the gunfire of our own vessels because their unlit presence at night represented a potential danger to the safe passage of convoys. At one time there were three abandoned cargo vessels floating upright on the sea waiting for the assistance which could not be provided. In the face of such evidence it became a part of major policy to retrieve every salvable vessel regardless of cost.

As a first step, the Admiralty of the day sought the assistance of present and past commercial tug operators, but because they had, only very recently, gone out of the business, the number of sea-going tugs which they could provide was, in the face of the problem, quite ludicrous. Even worse, the quality of this miserable offering was quite humiliating. It is perhaps pertinent to mention, in view of the magnitude of the problem, that the Admiralty thought it worth while to purchase, through the agency of a private operator, *one single ocean-going tug* from a neutral source. The only practical solution to the problem was for the Royal Navy to design a suitable range of tugs, cause them to be built, and then to train the personnel and man them themselves. This project was put in hand forthwith and prosecuted with vigour . . . a little too late of course, but nevertheless . . . with vigour.

The nation's shipbuilding facilities were however strained to the limit in making good mercantile losses and in providing the naval ships to fight the war, not to mention the task of maintaining the tremendous fleets of ships, both naval and mercantile which were already in being. So, the new tugs were slow in coming from the yards, and those completed, although vastly superior to anything then in service, were very far from being entirely satisfactory. The earlier arrivals, none the less, performed yeoman service, and their crews, with the experience so readily obtainable, provided the nucleus of the fleet which the situation demanded.

The Admiralty conception of the ocean-going tug-type was very fairly represented by the Saint Class. Tugs of this class were still being delivered from the building yards up until two years after the end of the war for which they were intended. After Admiralty had retained those units deemed essential to their own requirement, the surplus was disposed of to various commercial towing interests, to Commonwealth and Foreign Navies, to Empire Port Authorities and the larger Oil Companies. It was ironical that when tugs were needed there were none to be found, whilst after the crisis, ocean-going tugs were a shilling a dozen.

The Saint Class of tug, in the hands of these separate interests was developed from a somewhat disappointing, although reasonably satisfactory vessel into a type of some significance. The principal defects of the original design viz., poor endurance and indifferent stability characteristics, were remedied at one stroke by converting them over from coal firing to fuel oil. A side effect improvement also resulting from this major modification was the hull stiffening and increase in displacement tonnage which resulted from the physical modifications to the hull interior. Some units were equipped with elaborate fire fighting appliances and others were fitted with automatic towing winches. All of the more satisfactory conversions included the removal of superfluous top-weight and detail improvements to deck, engine-room, and domestic arrangements.

In considering the evolution of the ocean-going tug type from the British point of view, it is perhaps unfortunate that this class of tug responded so well to intelligent conversion seeing that their very continued existence stultified completely any further design development of the type in this country for nearly twenty years. The fact that these vessels were available in such numbers, and at such a low price, naturally precluded new thinking and fresh design and expenditure. This is deplorable but is demonstrably human nature.

British shipyards, which specialised in tug construction, did, of course, build a number of deep sea-going tugs during the years between the wars, but very few of these were bona-fide Ocean-going tugs, and even so, hardly any of them subsequently wore the Red, White or Blue Ensigns. Some of our domestic firms did build certain tugs which were capable of making ocean passages, but even so, their prime function was patently dock and harbour towage with a very limited open seas potential.

In Holland the situation had been regarded, throughout the period under review, from a different viewpoint entirely. Dutch interests had passed through the seeking phase to salvage with direction and intention. They had specialised in the business and had, in the formative years, recognised the very special place which must always accrue to the ardent specialist. They made a study of the problems of towing large inanimate objects by sea over vast distances and had become experts in the contract towage of floating cranes, craft without power, barges, dredgers, harbour construction equipment and so on. Although the World War I had quite naturally interfered with Dutch maritime activities, the resumption of peacetime operating saw them back, even more firmly, in their erstwhile position of pre-eminence in the realms of ocean towage. When it became necessary to tow a huge floating dry dock from the United Kingdom to Singapore, the Admiralty were very pleased indeed to entrust the operation to Dutch tugs and Dutch seamen. In fact, by the middle 1930's, the terms 'Holland,' 'Efficiency' and 'Towing Know-How,' were all three quite synonomous.

While Britons were busily converting the 'Saints,' the Dutch were involved with drawing board and test tank, and in 1933 they launched an ocean-going tug which was to set the pattern of open-seas work for the next three decades. This vessel of nearly 1000 tons gross, with her twin diesel engines geared to one shaft to produce more than 4,000 horse-power, with her range of 20,000 miles and her superlative sea-keeping qualities, was only surpassed in her obvious excellence for the work for which she was designed, by one other product of Dutch seafaring genius, this was the officer who was selected to command her. The Dutch tug *Zwarte Zee*, under the command of Captain Teumis Vet was to set the standards for absolute excellence in ocean salvage; standards which will remain unequalled for many, many years yet to come.

When 1939 saw the United Kingdom facing the old enemy once more, the Admiralty, although haunted by a thousand and one difficulties and complications, profited from the experiences of the First World War and ordered a series of tugs from a famous East Coast shipyard. The series was not limited in number, the order was . . . Build them and keep on building them . . . modify them as experience dictates but do not pause in the production. Until these vessels began to come into service in mid-1940, the Royal Navy formed an ocean-going tug section out of the craft available.

About a nucleus provided by three fleet tugs of the 'Marauder' class, the Navy assembled a heterogenous collection of converted and unconverted 'Saints,' a couple of reconstituted 'Rollicker' class of Dockyard tugs, a couple of American tugs which were, in the ensuing years of war, to obtain to all of their Commanding Officers the some-time sobriquet of 'Tug-boat Annie,' plus the pick of the commercial craft available. Of these, only those obtained from the Overseas Towage and Salvage Company, of which more later, could be really regarded as effective in the open-ocean capacity. This fleet bore the brunt of the results of the first onslaught until the new craft came in mid-1940 months; and leaving out those which went ashore the short way, namely vertically downwards, they continued, even with the new tonnage, until the very end, such were the demands upon ocean-going tugs in the years 1939 to 1946.

The new class of tugs were eminently fine vessels, their sea-keeping qualities were excellent and their design, although somewhat unimaginative and more inclined towards an enlargement of the harbour type of tug than to a specialised ocean-going tug, showed a considerable advance upon any previous British construction. The hull design was particularly good, although some aspects of the deck layout showed an imperfect appreciation of the practical aspects of sea towage. Their principal defects lay however in their quite inadequate endurance and their lack of power for heavy protracted towage. These defects were, of course, attributable to the fact that this class of tug was obliged, by the restrictions imposed by total war, to be steam powered. The merits of diesel propulsion were well known and fully appreciated by all concerned, but at that time such units were simply unprocurable. This class of tug was followed by a larger class of steam tug which had reduced, if not removed, defects of a like nature.

The real solution to the war-time sea rescue problem was not obtained until 1942, when the availability of suitable diesel engines permitted the production of a class of seven tugs which were the finest that Great Britain had ever produced and certainly the best that the world had ever seen. The Admiralty conception of a heavy diesel tug was a magnificent vessel of 1,100 tons gross with an overall length in excess of 200 ft. and a beam of 40 ft. and a loaded draft of over 18 ft. The main engines, totalling more than 4,000 h.p., were arranged to drive a single large propeller through a fluid coupling. Bunker space allowed of an endurance of more than 20,000 miles at full power. The hull design gave excellent sea-keeping qualities under the worst of sea conditions, even at speed.

The tug's equipment included a massive automatic towing winch, a special shock absorbing towing hook, a powerful capstan and windlass, a comprehensive arrangement of alongside towing bitts and a tremendous hold capacity for towing cordage and spare gear, etc. Besides fixed salvage pumps and air compressors, this class of tug carried a selection of portable salvage pumps and portable pipes and hoses, there was also the fullest fire-fighting equipment conceivable. These vessels rendered yeoman service in the Atlantic Ocean during winter months and elsewhere for the balance of the years of sea warfare. There are still six of these vessels in service, some with the Royal Navy, others with commercial interests under charter.

It would be very unfair indeed to proceed with any further commentary upon the evolution of this chosen ship type, without explaining that the three classes of vessel mentioned as new construction during the war years were, in fact, developments of designs originally prepared to the orders of the Overseas Towage and Salvage Company of London. The first class was a development of the s.t. *Salvonia*; the larger steam type was a development of the s.t. *Neptunia*, whilst the *Bustlers* were a development of a tug which that company had progressed to the design stage in reply to the *Zwarte Zee*.

With the almost unlimited experience which unrestricted U-boat warfare was providing, our Naval tug crews were rapidly achieving pre-eminence with the Dutch at the work. With magnificent tools for the job, and with the interest which success brings, our Commanding Officers were examining with expert eyes this new craft in which they had won their expertise, and their subsequent contribution shows in improvements and innovations to tugs, their equipment and to the line of thinking in towing operations and procedures. It was whilst they were acquiring this expertise that they discovered a fifth essential quality to accompany the original four designated for the seeking tugs of nearly a century before. They found that besides Size and Seaworthiness, Endurance, Power and Efficiency plus economy, the factor of pure *weight* must be added.

As the war progressed, the tugs were called upon to tow the very largest classes of vessel as they fell victim to the torpedo, the mine and the bomb. In calm weather it was found that once a towed vessel was persuaded to gather way, the maintenance of momentum was not too difficult, always provided that the tow could be kept under control and following well. In heavy weather however, or if the tow yawed at all, even in quieter weather, there was a tendency, even when the largest tugs were involved, for the tow to take charge. Upon examination it was appreciated that in addition to the usual heavy discrepancy between the displacements of

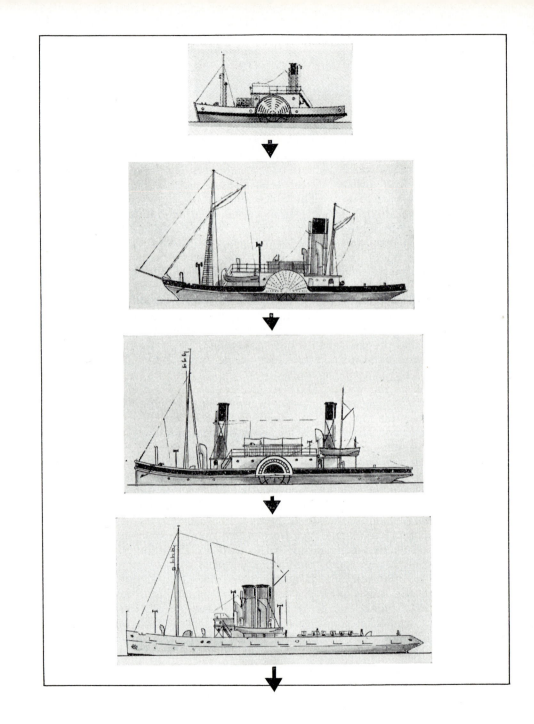

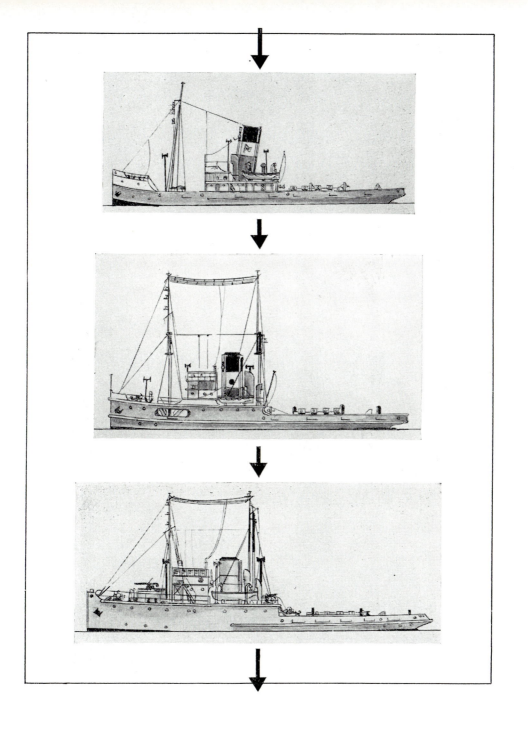

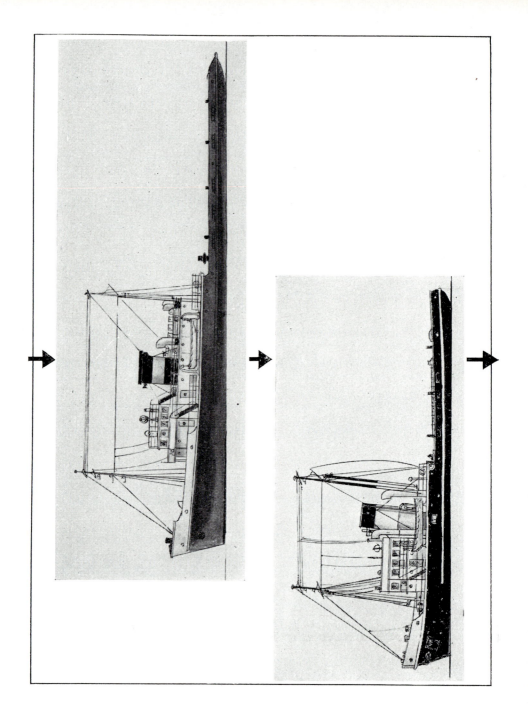

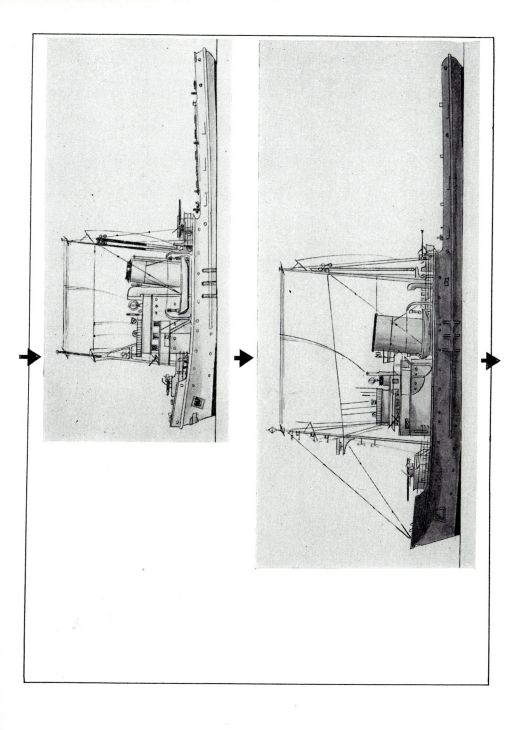

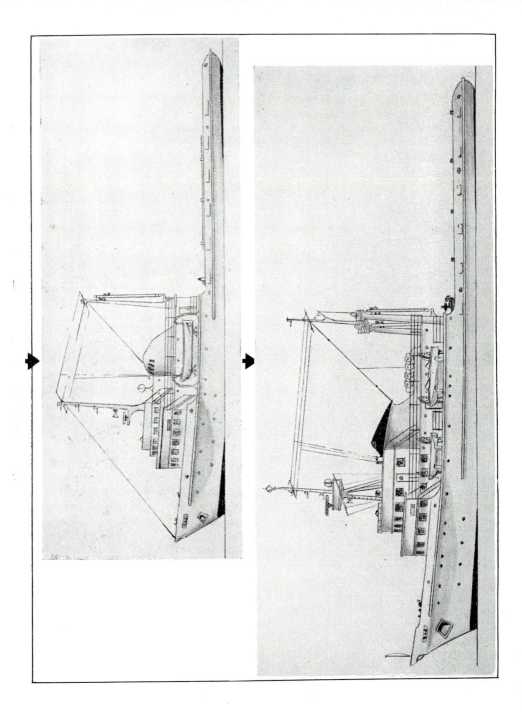

the tug and tow, consideration had to be given to the proportionate dis-crepancies in the above and below the waterline areas offered to wind and sea respectively. Unless there was an adequacy of displacement in the tug, it was found, her inertia, even when assisted by a relatively high propulsive force plus rudder action, could not fully control a hulk in tow. In the particular case of the tugs at sea in war, the lesson was to keep the tug in the fully laden condition by filling ballast and peak tanks, and by fiillng empty fuel tanks with sea water. In the general case of overall lessons to be learned, there were two. One was that the interior tankage of tugs should be deliberately arranged so as to provide the maximum displacement as required for towing. The second was that there was only the same limit applying to tug construction as applied to other ship types . . . in other words, as ships get bigger, salvage tugs must grow with them if they are to adequately serve the industry.

Whilst this nation had, in the shape of the 'Bustler' class of tug and the men who manned them, made a signal contribution to tug design and towing knowledge, the cessation of hostilities found us back in the doldrums once more. The surplus wartime ocean-going tugs found their way on to the commercial market at knock down prices, so that a repetition of the twenties once again obtained where operators imposed modification upon modification upon a design that was, in fact, obsolete as soon as the war ended. All operators that is except the Dutch operators. After a brief moment of appraisal, the Dutchmen screwed up their courage, drew in their belts and realised the last Guilder of their credit and built a brand new fleet of ships. Tug after tug, each embodying the experience of a century topped by seven years of the most intensive salvage experience in history. Ship after ship, each a little bit better than the last. Each tug a little bit more beautiful than the last. Tugs of from 1,500 h.p. up to 4,000 h.p. Each of these vessels was diesel powered of course; where the power required indicated the need, the engines were twinned but drove a single propeller. As the size of tugs grew steps were taken to enhance handling by particular attention to rudder design. Upon occasion resort was made to twin or box rudders where this seemed advisable.

With the wealth of experience available the Dutch decided that the demands of contract towage having indicated a commercial requirement for the towage of two and three objects simultaneously by one tug, then the tug's equipment must provide the desired flexibility, accordingly certain of their new units were fitted with a duplex towing winch and towing bollards. Man made fibres having proved adequate to the requirements of ocean towing, all of the new tugs are being supplied with a selection of appropriately dimensioned nylon hawsers.

Each new unit added to the fleet was equipped, of course, with a comprehensive array of electronic equipment for navigation and communication. All the while the Dutchmen were building this new fleet of tugs they were pondering upon the ultimate lesson of the war. The axiom of power plus weight and considering the latest application of that axiom. In view of the introduction of super-carriers of up to 200,000 tons deadweight was a new super-tug necessary? Could the expensive commitment of a super-tug be avoided by the use of a complex of more familiar sizes? The answer to that question brings this commentary upon the evolution up to date, because the answer was *yes*. The pioneers of the supertug of the 1930's have once more given a lead with a new vessel which almost doubles the power and size of her immediate forbear and does so in a shape which must surely represent the ultimate in beauty of line besides the ultimate in power and seaworthiness. (Figure No. 2).

SECTION 2.

CHAPTER I.

The Proportions and Design Characteristics of the Ocean-Going Tug Type

Whilst there will always be detail differences as between one naval architect and another as to the design proportions of ocean-going tugs, and whilst various nations' concepts of the type will also vary from one to another, it is a fact that the requirements as to basic proportions do not vary to any large extent. Looking at units of the type as produced by a selection of Northern European yards it would appear that a fair average L/D proportion obtains somewhere between the factors of 9·5 and 10, with a figure of between 4·75 and 4·95 representing typical L/B ratios. A freeboard ratio approximating to ·1B seems also to be fairly representative for contemporary application. Most naval architects seem agreed upon the necessity for introducing a measure of rake, rise, or trail into the keel-line in order to ensure of an effective immersion of propeller tips under all service conditions, and an average estimation for this consideration would appear to be close to ·04L.

If these design proportions are related to a tug of a proposed length of 200 ft., a beam of 41 ft., or thereabouts, is offered; a moulded depth 20·5 ft. is provided which, with a freeboard of 4 ft., allows of a midships mean draft of about 16·5 ft. When the rake, rise or trail allowance is applied to this latter value a maximum after draught of about 18·5 ft. resolves so that the type concept shows as short, beamy and relatively deep. Because the designed B proportion, in most modern versions of the type, only obtains for a short distance about the mid-length of the vessel, the resulting coefficient of displacement is quite fine.

For eminently practical reasons the free running speed of this class of vessel must be high; a speed of sixteen knots, for instance, would not be regarded as unreasonable or remarkable for a unit of 200 ft. length. The V/L relationship is therefore high, higher indeed than in most classes of merchant ship. This circumstance combined with the design feature of a fine coefficient of displacement, already referred to, has the effect of producing a considerable bow wave when the type proceeds at full running speed. Without specific design compensation the type is therefore somewhat vulnerable to heavy water taken in over the fore-ship when operating under conditions of adverse weather.

The design compensations which have been developed now provide the two principal visual characteristics of the type.

The first design compensation considered was a marked increase of deck sheer. This innovation returned an immediate measure of relief but produced a complication at the after part of the ship where the increased sheer set up an intensification of chafe and damage to towing media. The remedy was then modified so that the increased sheer was only applied to forward part of the main deck. This modification passed through a series of experimental phases before practice indicated maximum relief when fore-sheer was arranged at four times the amount of after sheer. (Figure 3).

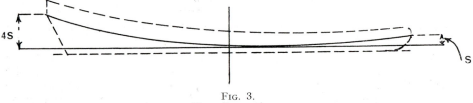

FIG. 3.
Sheer Proportions.

This modification was not however, by itself, the whole solution; this was ultimately provided by superimposing a raised, or top-gallant fore-castle head over the whole fore ship. As initially introduced at the end of the last century, this feature took the form of a triangularly shaped super-structure imposed over the fore end of the tug at main deck level. In this most primitive form it could be, and frequently was, added to existing tonnage as an uncomplicated modification. Even in this form when it really constituted no more than an ultra substantial weather-cloth, the value of the feature was instantly recognised and during the decade or so which followed, this feature progressed from a relatively flimsy accessory into a properly integrated design characteristic.

As arranged in modern construction this topgallant forecastle head extends over about one half of the vessel's length from forward so as to protect all access apertures to the accommodation and machinery spaces, it is therefore entirely practicable to design units of this type so as to provide passage from the Wheelhouse down to any position, or compartment, below the main deck without emerging into the open. It must be remarked that the forecastle head is integral with the hull proper. It is not a structure imposed upon the hull at main deck level, but is obtained by extending main framing up to the desired height over the required length at full strength and standard spacing with side plating to the full rule weight. It is usual to provide a beam at every frame station, with a deep knee at every third frame below the main deck. Both the main deck and the forecastle deck are of rule thickness and are sheathed with 2 in. teak or $2\frac{1}{2}$ in. pitchpine. Whilst sheer to both of these decks is arranged as has been already described, it is customary to half the sheer in the first deck over the forecastle head and to reduce it further at subsequent levels, this in the interests of the convenience and comfort of watchkeepers.

The rest of the construction of the type follows the orthodox transverse application, except that every naval architect, almost without exception, calls for a pair of heavy under-deck girders to run longitudinally from the after engine-room or cross bunker bulkhead right aft, in order to maintain full longitudinal strength. These girders are commonly arranged to coincide with the longitudinal boundary material to any substantial deck opening in the main deck such as the Engine-Room Skylight or the Salvage Store Hatch. Framing is rather closer than in comparative merchant types, size for size, and certain naval architects interested in the type call for even closer frame spacing in the fore ship because of the arduous nature of the function; one eminent specialist reduces frame spacing here by 18 per cent. The more normal subdivision of this type of ship, as required by tankage, etc., provides a hull of tremendous strength and great rigidity. Besides the fore and aft dividing material, average requirements will entail five oil or watertight bulkheads extending up to main deck level with another six or seven to the level of the lower deck.

Unlike the usual run of commercial construction, there is neither the need nor the desire, in tug work, to produce a good deadweight lift displacement tonnage ratio by designing for comparative lightness. Rather is the reverse the case; weight is a very necessary adjunct to towing power so that if the desire for a robust hull requires the implementation of heavy metal there is no design complaint. It is then customary to adhere to the classification societies' requirements in the matter of plating, bulkheads and the like but to increase dimensions where experience indicates the need. It is common to find rule depth for frames exceeded in tugs whilst naval architects frequently call for additional thickness in sheer strakes, at deck level at the ship's side and at, and about, the point of tow and in the stem and after rail vicinities.

The construction detail of ocean-going tug stems has excited a fair measure of contention between interested parties over recent years. One school holds firmly to the view that ocean tugs should never be used in any role other than pure deep-sea towage. Another school contends that it is quite inevitable that the type will, sooner or later, be required to work at close quarters when the ability to push may perhaps prove to be of equal importance to pure towage. The former school prefers to incorporate the modern soft plated stem into its designs whilst the latter use a rectangularly sectioned bar stem raked and knuckled at a point mid-way between the main and topgallant forecastle decks and backed with a very heavy breast-hook. The vertical face presented above the knuckle is then cleaded with permanent timber fendering or alternatively, according to the inclinations of either operator or naval architect, provided with eyebolts for the suspension of a suitable portable fibre, rubber or composition fender.

Interests in the development of the fore ends of ocean-going tugs has, of recent years, been very properly balanced by a growing concern with the after parts. This is most essential because of the large and complicated stresses generated hereabouts by a combination of the downwards trend of

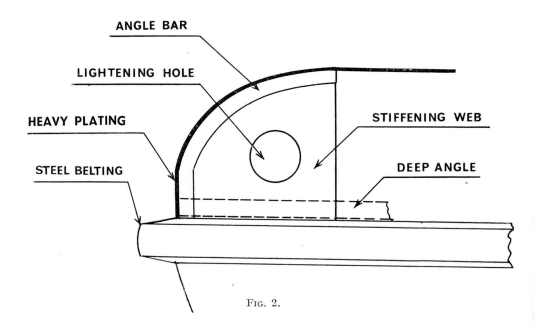

Fig. 2.

the towing gear from the after parts of tugs, the very considerable weight of the medium itself, and the even more considerable towing effort which must be transmitted. All of this causes friction of an extremely high order about a tug's after parts, particularly at the taff-rail, a circumstance which careful design can ameliorate, even if it cannot be wholly eradicated.

In their consideration of this problem, the naval architects are at pains to provide a contour profile for the more afterly sections of the bulwarking which will reduce, as far as is practicable, the prime source of the friction and will also carefully assess the weight of material required, and the type and style of stiffening indicated, in order to provide bulwarking which will properly support and withstand the combination of weight and stress which the function involves. (Figure No. 4).

Practically all classes of ocean-going tugs have hitherto been designed about the 'bar' type of keel. Prime fore and afters of this type have always been traditionally associated with tug construction because of the great longitudinal strength which they are reckoned to contribute to vessels which are, by the very nature of their function, obliged to suffer a very great deal of punishment for the whole of their working lives. This type of keel has moreover the reputation of enhancing a tug's course-keeping qualities whilst towing and is generally held to be a highly desirable design feature in vessels which must accept grounding, either by accident or design, as an occasional functional hazard. It is also the truth that it is both an expensive and tedious affair to fit intercostal keelsons over flat plate keels in vessels where the rise of floor is steep and where frame spacing is small. In such instance then it would appear that the demands of economy and efficiency walk hand in hand, at least in all but the larger classes of ocean-going tugs.

When tugs grow up to a designed length of 200 ft. and upwards, however, much of the difficulties attendant upon the installation of a flat plate keel, and its associated components, are removed and a very large school of opinion avers that the larger classes of tug should be so fitted. This school of opinion seriously contends that the handling characteristics of the larger tugs have not advanced pro-rata with other design and equipment improvements and that this may be attributed to the retention of the bar keel with its excess of deadwood both forward and aft, but more especially aft. It is argued too that the replacement of the traditional rudder of double plate construction by a fully balanced rudder, in association with the style of underwater profile which a flat plate type of keel would allow, would provide an enhancement of handling characteristics sufficient to immeasurably improve the flexibility of employment of these larger classes of tug.

CHAPTER II.

The Point of Tow in Ocean Tugs and the Appliances Located thereat and elsewhere for the purpose of Transmitting Towage Effort

A consideration of the very greatest importance in the design and construction of any tug must obviously be the selection of the point from which the vessel will transmit propulsive effort to any other vessel or floating object to be towed. This point will only be established after the most careful thought has been given to all of the relevant data, and to practical towage experience in the appropriate towage sphere, because the correctness, or otherwise, of the location selected will establish the tug's handling characteristics, and possibly even her very seaworthiness, for every type and condition of her subsequent towage employment. The position which is ultimately selected will, in all but certain classes of tugs employed in closed waters, always be something of a compromise but in the ocean-going classes every care must be taken in the proper assessment of the compromise in view of the arduous and hazardous nature of ocean salvage employment and, of course, the very high financial stakes involved.

Ideally, a point of tow would be selected so that the towage effort could be transmitted in such a manner as to offer the least possible interference with the tug's freedom of movement, in which case the point of tow would be at the tug's turning point. This location, or at least a point very close to it, is always chosen in the case of tugs working in sheltered waters where there will be no substantial seaway motion and where superstructure requirements are small. For a number of reasons the location of the point of tow precisely upon the turning point is not practicable, even with harbour craft, but one of the leading naval architects interested in tugs has successfully located the towing point in several of his deisgns in a position ·02L abaft amidships. Towage from this point gives a light but positive reaction to helm effect but offers no appreciable adverse reaction. The tugs involved were, of course, harbour tugs where circumstances permitted an unfettered selection of location, the conditions of towage considered were moreover those associated with close quarters work with the towing medium at short stay and trending upwards from tug to tow.

26

It is, however, entirely apparent that it is desirable, if not essential, to locate the point of tow as near as circumstances will allow to the turning point, and simple mechanics will clearly show that the farther one departs afterwards along the centre line from this point, to establish the vital location, the greater will become the resistance offered to the tug's turning efficiency by the resistance of the vessel or object towed to forward movement, a maximum effect of almost complete binding to large alterations of course being reached when the point of tow locates at the aftermost parts of the tug. The maxima and minima are therefore most clearly defined.

Whilst the location of the towing point must be selected with a very full consideration for its effect upon the tug's manoeuvrability, there are, of course, other conditioning factors of great importance and listed very briefly these are:—

1. The historically proven requirement of a topgallant forecastle head over the forward half of the vessel.

2. The desirability, in ocean-going tugs, of reducing the inboard scope of towing medium to a minimum in view of chafe complications.

3. The necessity for providing an effective platform for the stowage and operation of lifeboats, working boats and possibly firefighting or other salvage equipment.

4. The essentiality, in craft designed to operate in all weathers in the open seas, of providing effective protection to all apertures and accesses to below decks.

In any consideration of these and allied circumstances it must be recalled that when all ocean-going tugs were steam propelled, the mass of the engine and boiler assembly occupied about one half of the tug's length about amidships. The boiler space was usually arranged just forward of amidships and was enclosed within the topgallant forecastle superstructure whilst the engine-room proper was disposed abaft of this and below the main or towing deck, access, protection and ventilation being provided through a massive engine-room skylight occupying almost the whole of the towing deck and constructed to a height of four feet, or more in the larger classes, on the centre line. Then there was little or no choice in the longitudinal location of a point of tow, indeed its vertical location was likewise restricted. It was invariably arranged upon the after bulkhead of the boiler casing in a position which, depending upon the number of boilers fitted, and their disposition, could be anything up to ·25L abaft of amidships.

When the internal combustion engine was introduced into ocean-going tug design considerations the first benefit to derive, after the operational advantages had been assessed, was the design freedom permitted by the removal of the boilers and their attendant tankage and equipment. Some naval architects elected to install the new style engine in the space hitherto occupied by steam engine and retained the skylight with the practical inconveniences involved in order to shorten the forecastle superstructure to achieve a more forwardly point of tow, this, of course, in the search for improved manoeuvrability. Others speedily removed the main engines into the space vacated by the boilers and removed the engine-room skylight a deck higher; claiming thereby a vast improvement in seaworthiness and the provision of a literally unobstructed towing deck. In both cases however it is possible to provide a more satisfactory point of tow than obtained in the case of many of the steamers. A point $\cdot 1L$ to $\cdot 15L$ abaft of amidships is entirely practicable, even in the consideration of items 1 to 4, listed above, with main engines in the afterly location, whilst a situation $\cdot 2L$ abaft amidships is achievable in the case of engines in the more forward position. The employment of two or more engines to drive one propeller shaft, with the reduced engine heights thus permitted, has done much to provide further flexibility in this direction.

In assessing these alternatives, and the practical benefits deriving, it must be remembered that whilst the ocean-going tug is not generally expected to carry out the more complicated manoeuvres demanded of her smaller sisters, and can therefore accept a more afterly location for the towing point, with the diminished flexibility which is implied, an ocean tug's value to the operator will be much enhanced if she can retain the charge of her tows from the actual commencement of operations right to the end of her voyages, thereby reducing to a minimum and diminution to her earnings as represented by the undue employment of harbour tugs. On the other hand, if improved manoeuvrability, obtained by the selection of a forwardly point of tow, occasions additional chafe to the towing media, or if the presence of a very large engine room skylight upon the after deck reduces, to any appreciable extent, the seaworthiness of the tug then it is possible that any marginal gains won could well be wholly cancelled through the loss or abandonment of a tow or even by the loss of the tug herself. Hence the introductory paragraph's insistence upon the care necessary in the correct appreciation of the various aspects of compromise.

The location of the towing point in the vertical plane, and relative to deck levels and to any tug's centre of gravity, is relatively easy of assessment seeing that it is almost wholly conditioned by purely practical factors:—

1. The complete integration of the point into major construction members in view of the stresses to be accepted.

2. The preservation of maximum stability under all conditions of operational usage by reducing the lever, as represented by the vertical distance between the point of tow and the tug's C. of G., to an absolute minimum.

3. In connection with 2 above. The elevation of the point, however, to such a height as will reduce inboard chafe to the towing media to a practical minimum.

4. Rendering the point fully accessible to the tug personnel at all times and in all conditions.

The conditions as set out in items 1, 3 and 4 above are clearly readily obtainable by selecting a position as close to main deck, level as may be practicable; the condition at 3 providing the only serious complication to this aspect of location. If the significance of this condition is to be fully appreciated a full understanding is necessary regarding the lead of towing media, both within the confines of the tug's fabric and thereafter. It is also desirable that there should be a full appreciation of the scope and variety of the stresses required to be absorbed by the towing media and their associated appliances.

When a tug is towing at short stay the tow rope lead is a straight line from the point of tow to the first bearing point upon the vessel or object towed. Because practically all vessels, and most towable objects, float higher in the water than tugs, the trend of the tow rope is almost invariably upward and the sole point of contact, and only sources of friction within the tug, reside at the actual point of tow. As the scope of gear in use increases, the weight of the medium itself asserts and a sag or catenary develops between the tug and her tow so that in towing any object of any appreciable substance with any appreciable scope of gear this sag, or catenary, will develop to such extent as to immerse the greater part of the towing medium in the sea. The tow rope trend will therefore be almost invariably downward from the after parts of the tug whilst towing at sea.

This downward trend will rarely be constant, even in the most clement weather, seeing that it will change with every yaw and surge of the vessel or object towed. Under conditions of adverse weather the separate movements of tug and tow will aggravate this condition so that in violent weather the sag may momentarily depart entirely; the tow rope may even occasionally assume an upward trend for brief periods when the tug pitches heavily, but in such case it will almost certainly be followed by an almost vertically downwards trend before tug and tow resume a more normal relative position the one to the other. Average angles of dip in the gear from the tug's after rail will normally present at from 10° to 20° whilst absorbing a loading of no mean proportions.

In ordinary weather the stresses involved begin with the actual weight of gear in use which, even when wholly immersed, can present at between three and four tons, to which must be added the actual tow-rope tension which, for a very modest tramp type of vessel proceeding in still water, could provide an additional five tons of loading, thus providing a norm of say eight to nine tons. If weather conditions deteriorate, circumstances might easily develop so that the tug may well exert her full power without advancing the towing unit through the water so that total tension in the gear may resolve at up to fifty tons or more depending upon the tug's bollard pull. If conditions now obtain where the tug and tow encounter sea and swell with vastly differing roll and pitch characteristics, then the stress imposed upon the towing medium and its associated components may well be away in excess of bollard pull plus the weight of the gear, and be quite incalculable, especially when the tug may be traversing the face of a heavy swell whilst the tow is encountering the full force of a breaking head sea. Under this latter circumstance, if the duration of the loading persists for longer than the period required to take up the extensibility provided by the catenary in the gear, then an immediate fracture of the towing medium is inevitable.

If the stresses involved are now obvious, it must be equally obvious that damage to towing media resulting from chafe deriving from such sources can be both dangerous and extensive seeing that it must persist over the whole scope of medium lying within the confines of the tug herself, primarily of course at the taff rail, but also over the towing horses and at the point of tow whilst the lead of the gear runs truly fore and aft, but clearly becoming more complicated and difficult of prevention if it is constrained to run out over the bulwarks at any consequential angle to the true fore and aft line.

The Figure No. 5 endeavours to portray this problem. The after profile chosen is a fairly representative type and only major items of equipment are included seeing that these are the only ones to affect the towing medium and its lead. The illustration shows that no vertical disposition of the point of tow which is practically obtainable can eradicate chafe risk. Even at the lesser angle of tow rope dip of ten degrees from the horizontal the elevation of towing point required to keep the gear entirely clear of the after rail is 10 ft. above the deck level and about 21 ft. above the tug's probable centre of gravity. At this elevation, and should the tow rope, for any reason, be permitted to lead abeam, the tug might well be subjected to a capsizing force well in excess of 2,000 foot tons before the medium fails. The illustration shows that even drastic modification to a tug's after profile can only reduce, but can never eradicate, chafe to towing media.

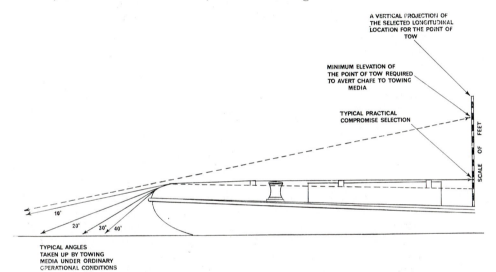

A VERTICAL PROJECTION OF THE SELECTED LONGITUDINAL LOCATION FOR THE POINT OF TOW

MINIMUM ELEVATION OF THE POINT OF TOW REQUIRED TO AVERT CHAFE TO TOWING MEDIA

TYPICAL PRACTICAL COMPROMISE SELECTION

SCALE OF FEET

10°
20°
30°
40°

TYPICAL ANGLES TAKEN UP BY TOWING MEDIA UNDER ORDINARY OPERATIONAL CONDITIONS

Fig. 5.

Considerations involved in the establishment of the Towing Point.

Seeing that the only practicable remedy to the incidence of heavy chafe lies in the elevation of the point of tow, and seeing that any substantial elevation of the point, to any degree likely to ameliorate the condition, only introduces problems of another character the only palliative measure open is to locate the point in such a fashion that it is not lower than the foremost bearing area on the towing horses and as little higher than bulwark level as may be consistent with other requirements. The compromise height reached by most naval architects is a point approximately four feet above main deck

level, the handling of the towing gear being facilitated, as and when necessary, by the provision of a sturdy table about the point at a suitable height.

Having considered the circumstances which condition the selection of
the point from which propulsive force may be transmitted to another ship or
floating object in tow, consideration may now be given to the methods used
to connect the towing medium to that point.   There are three ways of doing
this:—

1. By turning up the towing medium to towing bitts or bollards.
2. By passing a spliced eye formed in the towing medium over the bill
of a suitable hook fastened securely to the tug's fabric.
3. By means of a towing wire wound on to the drum of an electric or
steam towing winch.

Many ocean-going tugs are provided with two of these three facilities.
Tugs are frequently provided with a towing winch and either a towing hook
or a set of towing bitts whilst a few years ago it was common to see tugs
fitted with a set of bitts and a hook, but it is rare indeed to find a tug which
is provided with all three, principally because of the difficulties involved in
providing a fair run of the towing medium at a satisfactory level relative to
all three facilities.

In the pioneer days of towing, the method of securing the medium inboard
of the tug was determined by the prejudices of the operator, interpreted, to
the best of his ability, by the shipbuilder.   To some the logical solution was
to forge a substantial hook and to secure this hook to the tug in as robust a
fashion as ingenuity could contrive, the tow-rope being then secured thereto
by means of a seamanlike splice, adequately served for protection.   To
others the solution appeared as a sturdy post or bollard securely integrated
into the hull to which a tow rope might be secured by means of simple turns
and hitches.

The bollard school provided an answer which was the logical product
of accrued experience . . . 'Masts had propelled ships since time immemorial,
there was a wealth of experience immediately available concerning the
proper integration of masts into vessels' hulls: methods proven satisfactory
for the transmission of wind power to a hull must prove equally satisfactory
in transmitting propulsive power from one vessel to another.' . . . Accordingly
posts of mast-like girth were installed into tugs at the chosen point of tow.
Sailing ships' masts were stepped upon the keelson and partnered at the deck
to a suitable reinforcement by means of hardwood wedges, so this method

was adopted, the only difference being that the towing bollard was of square section up to the partnering in order to eliminate any rotation which may be imparted by hitches when under stress. Again, as with sailing ship practice, the stresses of propulsion were transmitted to main construction by laying an iron deck across the deck for the space of two or three beams on each side of the post. (Figure No. 6).

As tugs gathered power and weight and tackled larger tows the bollard fitting was modified to iron construction and then to steel. The advent of

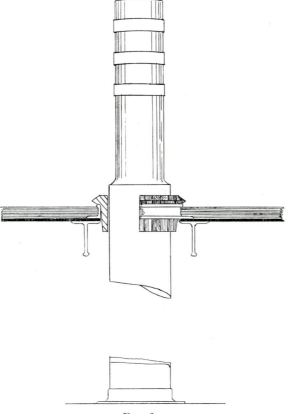

Fig. 6.
Typical early Towing Post Installation.

screw propulsion rendered the keelson heel location untenable and the twin legged cruciform bollard was adopted which forms the basis of today's appliance.

The contemporary version of the towing bollard is a massive construction in heavyweight seamless steel tubing which is usually arranged to occupy approximately one half of the tug's breadth about the amidships at the chosen point of tow. The twin tubes reeve through the main deck and extend down to the tank tops, or other major lateral element of construction, where they are secured by angle bar rings, usually hand-forged and fitted in view of their importance. Where the verticals pass through the deck beams are either curved or are laid flange about if the diameter exceeds the clear spacing. It is customary to lay a doubler above the plating in the way of a bollard and to secure the verticals to the doubler by means of another pair of forged angle bar rings or flat plate rings; extremely well fitted in either case.

The athwartship member of the bollard is secured to the foreside of the verticals by welding, suitably shaped holes being cut in the member to accept the circumference of the verticals up to a depth of one quarter of their diameter. Because this cutting process was unacceptable to certain operators, it has become the practice to construct the athwartships member of elliptical or bolster section so that when holes are cut for the purpose of reeving it over the verticals, the sectional area remaining is not less than the sectional area of the verticals. The horizontal member is then secured at the desired height by welding. All welding work is associated with towing appliances is required to be of the highest possible quality and is subjected to rigorous test and inspection before, and after, it is ground and buffed to offer the smoothest possible bearing surfaces to towing media. (Figure No. 7).

In order to facilitate the handling of heavy towing hawsers about these bollards, their dimensions, proportions and positions relative to other fittings are related to the largest size of hawser which the tug is likely to use. This is invariably the fibre deep-sea tow rope, and if the diameter of this is represented by 'D' then the minimum diameter of the verticals is represented by 3D. The horizontal distance between the inner faces of the verticals should be at least 12D with the extension at each side projecting to a distance represented by 6D. The upper side of the horizontal member is arranged at the same height as the upper surface of the forward towing horse and the verticals project above it to a height of 6D.

The thickness of metal used will vary with the tonnage and power of the tug but the larger classes will utilise metal of $\frac{1}{2}$ in. thickness and over.

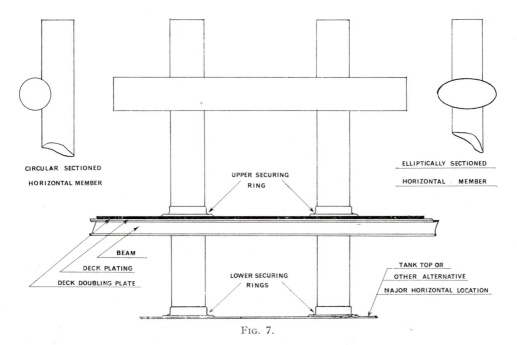

CIRCULAR SECTIONED
HORIZONTAL MEMBER

UPPER SECURING
RING

ELLIPTICALLY SECTIONED
HORIZONTAL MEMBER

BEAM

DECK PLATING

DECK DOUBLING PLATE

LOWER SECURING
RINGS

TANK TOP OR
OTHER ALTERNATIVE
MAJOR HORIZONTAL LOCATION

FIG. 7.

Coping with more than one tow simultaneously is entirely practicable with such a bollard but some operators consider that the flexibility of the principle is much enhanced if a third vertical is fitted. (Figure No. 8). The

FIG. 8.

additional vertical is usually arranged off-centre unless it is reasonably convenient to arrange a stool or other device over the shafting. The very largest classes of tug occasionally mount two complete towing bollards, side by side, at the point of tow to fully facilitate the securing of two or more sets of gear when concerned with multiple tows.

The original towing hooks were simple smith wrought hooks, either of the circular cup hook style or plain turned up shape. The circular style found favour mostly in the South of England whilst around on the West Coast the turned up style was more readily adopted so that in due course it became known as the 'Liverpool' hook. Both types were provided with substantial pads, in the first instance, and these were most carefully drilled to accept fitted bolts or rivets to provide the safest possible fastening to appropriate locations. A very short period of operational experience revealed a quite unanticipated high degree of frictional wear on the inner surfaces of the splices in the tow-ropes. This required the provision of swivelling facilities, both in the vertical and the horizontal planes.

This function was provided by two modifications. The first consisted of a traveller bar of suitable radius provided with palmed ends for effecting a connection to the tug's fabric by bolts or rivets, the hook proper being attached thereto by means of a shackle of suitable dimensions. Whilst this appliance was somewhat unsightly, its rugged simplicity promised trouble free service. Operators with a somewhat more advanced aesthetic requirement opted for an alternative mounting which consisted of a neat forged hinge block arranged to secure to a convenient vertical surface by means of through bolts or rivets. This block was drilled to accept a vertical fitted pin arranged to hold a swivel piece which in turn supported the hook.

Subsequent towage experience soon indicated the need for the rapid disengagement of the tug in emergencies. Following upon the practical suggestions of the tug's personnel, the constructors engaged themselves with the matter and in fairly short order there was a variety of slipping, or tripping, towing hooks of both the 'Circular' and the 'Liverpool' types available to operators. Most of the new devices were variations upon a simple principle whereby the hook proper was pivoted upon a robust horizontal pin in such a manner that the centre of gravity of the hook lay abaft of the pivot. The section of the hook assembly forward of the hook proper was fashioned so as to provide an accurately fitted 'flat' which could be engaged and held by an equally accurately finished latch to provide the assembly with an effectively secured closed position. This latch was, in turn, accommodated into a bifurcated forging introduced between the hook proper and the hinge block or traveller bar, depending upon the type. When the hook was to be tripped, the latch was moved forward, this released the fitted 'flat' and the hook swung downwards, under its own weight, to release the eye-splice in the tow rope. A safety device was subsequently arranged in the form of a stay arranged to fit over the nose of the hook and held in position by means of a pin fitted to a hole drilled transversely through the end of the hook.

With an improvement in the power, weight and size of tugs, tow ropes became heavier and the towing hooks were scaled up to accept them so that, in due course, the weight involved in an assembly of towing hook and swivel together with a spliced eye provided with a thimble and link, became a factor of some consequence, particularly when a tug provided lively motion in a seaway. It first of all became necessary to provide a bearing ring or table and to devise a method of holding the hook in the open position in order to conveniently accept the eye of the tow rope. The first requirement was adequately satisfied by common sense arrangements of plating and angle iron, but the second required a very careful consideration before the proper locations were established for holes and pins of dimensions suitable to the function of supporting the weight of the pivoting section of the hook and the eye of the tow rope. It later became the custom to drill these holes, and the ones associated with the tripping gear to the same diameter so that one pin, secured by a lanyard of adequate length, could be used to hold the hook in the open position until the tow-rope had been attached thereafter removing it to the nose position to secure the hook for towing. This provided a safety measure in so far as the loaded hook could not be inadvertently left in the open and held position. (Figures No.'s 9 and 10).

The desirability of incorporating some degree of resilience or shock absorbing medium into towing hook assemblies was appreciated at a very early stage in the evolution of the appliance. At first this improvement appeared to be very easily attained and a number of spring loaded hooks became available at very short order. The usual method adopted was to separate the hook assembly, forward of the tripping gear, into two parts one being an extension of the hook shank in rod section, the end being provided with a robust nut and washer to match a threaded length. The swivelling section of the assembly was extended for a similar length in the form of a cage or cylindrical housing of dimensions suitable for the acceptance of a large compression spring of a strength appropriate to the tug to be fitted. The hook shank was then inserted piston-wise through the spring, and its housing, through a retaining device attached to the housing, to be secured thereto by the nut and washer supplied. In this manner the spring absorbed

D

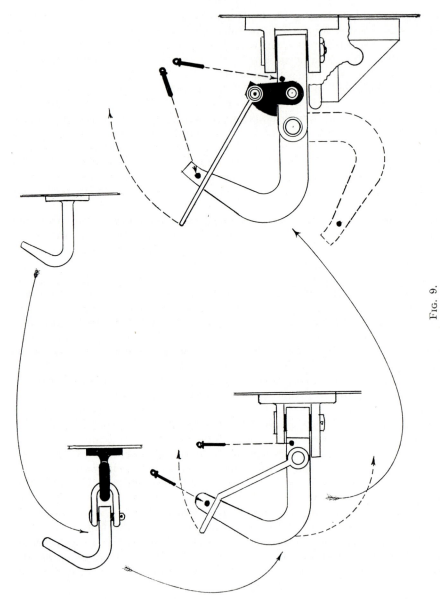

Fig. 9.

The development of the 'Liverpool' type Towing Hook.

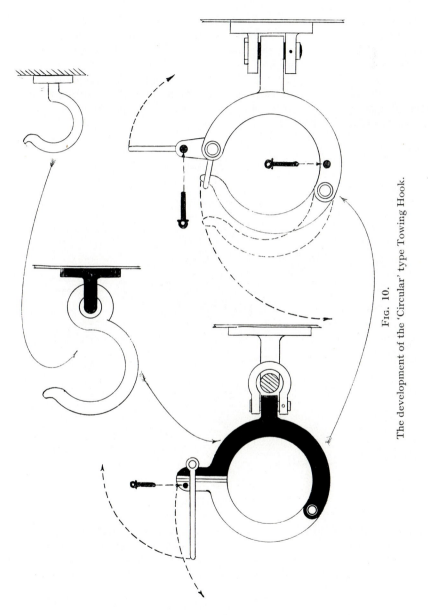

Fig. 10.

The development of the 'Circular' type Towing Hook.

the stresses of towage and was reasonably effective in dealing with shocks of short period such as arise when a tug first takes up the weight of a tow, or manoeuvres about with weight upon the hook, but the effectiveness of the device quite naturally vanished the moment that the spring was compressed fully. It was therefore clearly quite useless in any consideration of the stresses of high order and long period which must be provided for in any contemplation of shock absorbing equipment in relation to ocean long distance towage. (Figure No. 11).

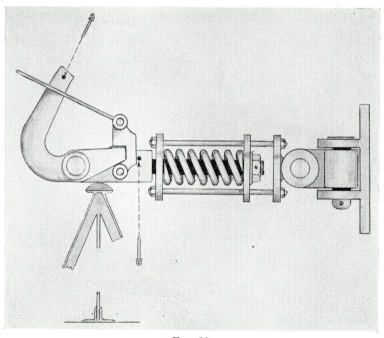

Fig. 11.
Early Shock Absorbing Towing Hook Assembly.

Because no solution to this problem was readily forthcoming, this setback led to a loss of interest in the project and Tug Masters were obliged to carry on as before, making every effort to avoid, by appropriate helm and engine movements such towage shocks as were avoidable but inevitably tow ropes continued to fail upon occasions for the simple reason that men tire after many hours of concentration, reflexes slow down so that appreciations become erroneous in the consideration of wind, sea and swell and the movements of the tug and her tow in heavy seas. The need for some measure of effective resilience at the towing point position was very real to the seamen and they looked forward, with no little anticipation, to some effective device which could relieve them of some of the wearisome vigilance demanded by the employment of rigid gear which depended solely upon the length and catenary of the towing medium for any measure of elasticity. The seamen were not, however, entirely alone in the consideration of this problem because academic interest had been stirred by this apparently insolvable problem. Part of this academic approach consisted of the preparation of fair statement of function which read as follows:—

1. To design a towing hook assembly so that it will cushion against shocks of short period but in such a fashion that although the resiliency embodied in the device is quite fully employed, there will always remain a sufficient reserve to accept the full tractive effort of the tug over and above the initial stress.

2. The shock absorbing medium must be fully effective over the whole range of towing stresses, from the full bollard pull in the heaviest weather, right down to the minimum effort required to promote steerage way upon a vessel floating in still water.

3. The shock absorbing medium must never compress into solidity even under the most extreme of stresses.

4. The device, if it were ever developed, must be rugged, easily maintained and serviced, and should be as nearly foolproof as may be contrived.

5. The device must be proof against the ravages of maritime exposure and occasional complete immersion in salt water.

6. Finally, any modification away from existing types of assembly must not be such as to reduce the manhandleability of the appliance under active seagoing conditions during adverse weather.

Although a simple compression spring, acting by itself, could not cope with the range and variety of shock and stress to be anticipated, it was offered, the spring could suffice if only it could be assisted by the leverage principle as exemplified in the steel yard type of weighing device. It was further offered that if the stress which was causing a compression of the spring could be automatically compensated by a system of levers which would increase the resistance of the spring to further compression, in proportion to the stress upon the towing medium, then all of the requirement of paragraphs 1 and 2 of the statement of function would be satisfied. In addition, if the leverage arrangements could be so devised that the leverage in favour of the spring, at extreme stress limits, could be such that the power of the spring multiplied by the effective lever length would give a resistance approximating to, or exceeding, the breaking strain of the towing medium, then all of the requirement at paragraph 5 would be realised seeing that, in such case, the shock absorbing medium could never become fully compressed even up to the maximum realisable limits. With an effective principle to work upon it was fairly obvious that the requirement of paragraphs 4 and 5 would fairly readily be obtained.

The sponsors of the new principle, following previous utilisations of the spring principle, designed his hook assembly in two sections, one part consisting of the hook proper, together with its tripping gear, the other comprising the hull attachment and the swivelling gear. These two sections were then fitted so that they would, at their point of connection slide, the one within the other. These sliding sections were then slotted to accommodate a pair of trunnions of dimensions related to the bollard pull of the tug; these trunnions were each fitted to carry a pair of large cams. These cams were, in turn, drilled at their upper ends to accept a second pair of trunnions, one pair being fitted to carry a rod arranged to pass freely through a bushed orifice in the others, the rod being extended to support a compression spring matched to the bollard pull of the tug. The cams were most carefully designed so that, at their upper ends, the profile curvature was such as to approximate to zero, the curvature being progressively increased to a maximum at the lower trunnion ends.

When this appliance was subjected to loading, the lower pair of trunnions being free to slide in the slotted guides provided, drew in towards one another and, because the initial curvature of the cams was slight, their point of contact . . . (which provided the fulcrum for the subsequent leverage upon the spring) . . . only moved very slightly downwards so that the resistance of the spring to compression was only very slightly enhanced, the spring flexed easily therefore under small stresses. If the loading on the appliance was then increased, suddenly or progressively, for a long period or a short one; then the trunnions, in taking up positions proportionate to the degree of loading, caused the cams to rotate to an appropriate degree to bring their bearing point upon sections of their profiles having a greater degree of curvature. This automatically increased the leverage assisting the spring so that balance was reached, and maintained, over the period that it was required.

To fulfil the difficult requirement that the spring should never compress into complete solidity, the cam curvature at the lower limits was deigned of such a radius that when the cams are brought into contact over this area, the leverage in favour of the spring is able to balance out forces greater than physical circumstances may produce so that the spring cannot, in fact, *EVER* be totally compressed.

Although this device clearly entails an increase in the weight and length of the towing assembly which, like all of the earlier devices of a like nature, demands the provision of a large bearing support, there is no inconvenience entailed which is not more than amply compensated by the benefits which it brings, and this type of towing hook assembly is an established favourite with the seamen and is included into the vast majority of new building specifications by most operators. (Figure No. 12).

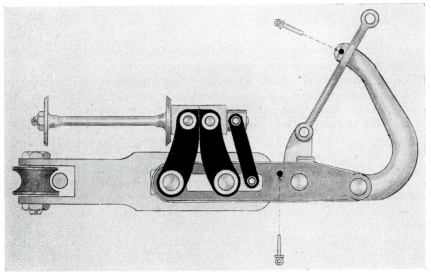

Fig. 12.

'Monarch' patent type Shock Absorbing Towing Hook Assembly (less spring).

When the ocean-going tug type is not fitted with a towing winch and when the towing hook is selected in preference to a towing bollard, it is customary to fit the hook assembly in duplicate so that the tug can cope with the towage of two vessels or other floating objects simultaneously. This presents no problem when the travelling bar type of swivelling gear is in use, but it does occasion careful design consideration when the hinge block provides the attachment. This is because when two sets of towing gear lie out abeam on the same side, the inboard towing hook assembly may bear very heavily on the outboard gear thus setting up awkward stresses in the area. Although this can be substantially reduced by careful fitting and the provision of suitably located buffers, it is generally held among tug men that the traveller bar, in this instance, offers the more trouble free service.

Because the towing medium will, under certain circumstances, lead out abeam of the tug whilst enduring strain of some consequence, tugs are, from time to time, subjected to large capsizing moments. Certain operators, in their appreciation of this hazard, have requested naval architects to provide the wherewithall for a movement of the towing point off the mid-line under these conditions. In the satisfaction of this desire to reduce the capsizing moment, some architects have fitted traveller bars of quite considerable radii, thus proportionally reducing leverage with the movement of the towing medium away from the fore and aft line. It must be observed, however, that to be effective, the traveller bar must be of considerable radius which must necessarily impede action upon the towing deck.

When the towing hook is the chosen towing appliance, the most common location for it is upon the after bulkhead to the top-gallant forecastle superstructure. The actual detail in regard to the means of attachment to this point, whether it be with a hinge block or with a traveller bar, will naturally occasion minor differences as, of course, will the preferences of individual operators, whilst the design considerations of different naval architects will provide evidence of previous construction experience. It is, however, evidence of a general soundness of application in that the general pattern does not show undue variation.

Almost all naval architects begin with asking for a substantial increase in the weight of plating used to form the after bulkhead proper. The deck in the area, the deck head above and the side bulkheads, for a little distance forward on each side, are also constructed of heavier than rule plating. The angle iron providing the boundary of the area is usually fitted at an increased depth, often twice on each flange. In association with this there should be, ideally, a complex of deep stiffeners to the after bulkhead together with brackets, etc. to properly integrate the strength required for the proper transmission of the stresses of towage down into main fabric, but space in tugs, like all other small vessels, is at something of a premium so that the same stiffening effect is usually achieved by subdividing the space forward of the point of tow into a number of useful small compartments such as lamp-lockers, paint or rope stores or deck W.C.'s, the sub-dividing metal being of the same weight as the main structure, adequately connected to the deck and to the deck head, and related to one another, and to the outer bulkheads by at least one horizontal member of substantial dimensions.

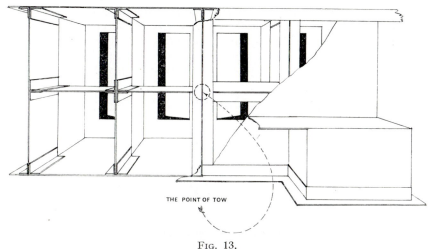

THE POINT OF TOW

FIG. 13.

Structural re-inforcement at the point of tow.

The after bulkhead is often buttressed, when convenient, by arranging bunker hatch coamings adjacently. Sometimes a pair of such hatches arranged abaft of the after bulkhead are bridged over to provide a convenient table for working about the towing hook. There will, quite clearly, be a number of alternatives available in this vicinity depending upon the ingenuity of the architect and the preferences of the operator, but the aim of all is to provide a massive and properly integrated cellular structure feature at this position. (Figure No. 13).

When, for any special reason, the after bulwark location is not available as a location for the point of tow, such as when a tug is fitted with a towing winch and the obvious position is occupied by winch fairleads or spooling gear, it becomes necessary to provide a separate stand for the towing hook. Such stands must obviously be of tremendously strong construction and must be most effectively worked into main hull members in view of the elementary stresses of high order which they must withstand. Where advantage may be derived from integration with prime below-decks members such as beams and bulkheads this must be done, but where this cannot be done auxiliary under-deck stiffening and strengthening must be provided. Where the stand is designed as an entirely separate component it is usually installed as a heavy tubular or box shaped structure extending from suitable major lateral construction below decks up through the main deck to a height compatable with all other relevant requirements. The connection at main deck level is usually provided by angle material of heavy section and deep flange, carefully fitted. Under the deck connection takes the form of welded bracketting arranged to integrate over three beam spaces. Such stands are

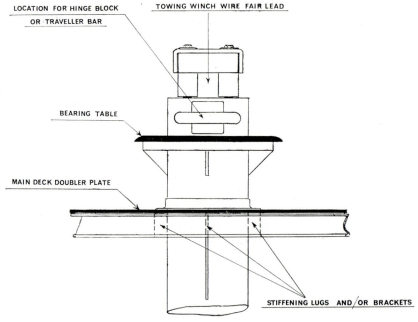

Fig. 14.

frequently elaborated to provide towing winch fairleads or a modified form of towing bollard, though, in strict fairness, it must be allowed that the latter modification rarely gives full satisfaction. (Figure No.'s 14 and 15).

In all modern ocean-going tugs both the towing hooks and the towing bollards have been relegated to something of a secondary role, being presently regarded as auxiliaries to the main towing appliance, i.e. the Automatic Towing Winch. The first automatic winches were steam powered, but little purpose is served in dwelling unduly upon the detail of their operating principles and design seeing that steam has wholly surrendered to electricity as the prime mover in this application. A later chapter will deal with towing winches proper, but at this juncture the intention is to indicate, in the most general of terms, the local stiffening and strenghtening which is associated with the installation of winches at the tugs' points of tow.

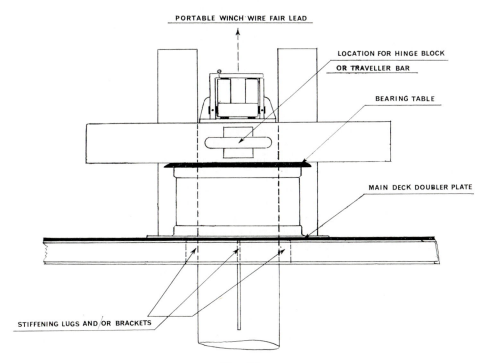

PORTABLE WINCH WIRE FAIR LEAD

LOCATION FOR HINGE BLOCK
OR TRAVELLER BAR

BEARING TABLE

MAIN DECK DOUBLER PLATE

STIFFENING LUGS AND/OR BRACKETS

Fig. 15.

In the consideration of automatic towing winch installations it must be remembered that, besides the stresses normally associated with towage, winches, and their associated features, are obliged to contend with the added effects of winching in a tow whilst towage is maintained. It must also be taken into consideration that a towing winch, together with its burden of 350 to 450 fathoms of very large wire, constitutes a gross weight of many tons, a substantial weight which will be concentrated over a comparatively small area.

As in every other aspect of tug design, all naval architects, and the operators themselves, have a most strongly developed and individualistic approach to the solution of the manifold problems involved. In the search for strength and rigidity, however, practically all of them demand a perfectly flat bed for the winch chassis, a condition which demands, of course, the total elimination of deck sheer and beam camber over the area to be occupied. Strategically located, substantial fore and aft under deck girders are also a practically universal demand, whilst a pair of substantial pillars are usually provided to act with these in the integration of strength and stiffness to main construction.

Deck plating in the immediate vicinity of the winch location is invariably provided at considerably in excess of rule thickness and experience has shown the wisdom of increasing deck plating thickness to some distance away from the winch location in every direction. When the winch is contained within a proper winch house, it is a common practice to augment the weight and section of materials all round, some designers calling for fore and aft girders, of appropriate weight and depth, to be fitted under the upper deck as well as to the main deck.

This chapter, has so far, only dealt with the arrangements provided for towage from the orthodox point of tow, but even the very largest of the ocean-going tug types is occasionally required to tow from positions alongside of other vessels or other inamimate floating objects requiring towage assistance. The ocean-going tug is also required, albeit upon extremely rare occasions, to provide steering assistance to disabled vessels. These two aspects of the towage function demand therefore the provision of effective appliances of appropriate strength, at correct locations, if the tug is to be capable of developing her full potential in assistance.

Whilst the ocean-going tug's normal equipment of mooring bitts and fairleads are normally provided at heavier weight and larger dimension than those fitted to an ordinary commercial vessel of like tonnage, it is still generally held to be advisable to provide such craft with at least one pair of heavy bitts on the forecastle head to cope with alongside tows. Some naval architects and operators hold that the provision of just one set of bitts forward is entirely adequate to the need in view of the fact that the orthodox appliance at the point of tow provides an efficient connecting point on the after deck. Others take the view that alongside towing bitts replace standard equipment and do not therefore seriously increase building costs, they also aver that . . . 'The unusual becomes commonplace in towage work' . . . and that a set of heavy towing bitts aft are bound to prove useful sooner or later. They therefore incline to the provision of two sets.

When towing bitts are provided forward and aft they are, of course, both fitted on the same side of the tug, the starboard side. This arrangement puts the tug upon the port side of the vessel or object towed and therefore on the inside when navigating upon the starboard side of a channel or fairway, thus allowing the tug's watchkeepers an unobstructed view of the oncoming traffic. In the ordinary way the standard bitts to be replaced by towing bitts will be the aftermost outboard pair on the starboard side of the fore-castle head and the forward outboard pair on the starboard side of the after deck.

Following modern ocean-going tug construction practice these alongside towing bitts are made up from seamless steel tubing of a diameter and section appropriate to the gear to be used and the power and weight of the tug. Twin tubes are rove through precisely cut holes in the deck and, seeing that these bitts are located, both forward and aft, in positions where the hull has considerable flare, they will be heeled upon stools fabricated from mild steel plate and worked into the framing and up to the shell plating by weld-ing. The actual connection to the stool is often provided by a closely worked angle bar ring whilst the under-deck connection is either similarly arranged or achieved by means of a flat plate ring according to the wishes of the architect.

Where these bitts are fitted to a sheathed deck it is the common practice to keep the sheathing well clear of the vicinity so as to allow the first turn of rope about the bitts to lay as flush to the steel deck as is possible.

The tubes are arranged at a proper distance apart bearing in mind the size of gear to be used and if whelps are fitted it is best practice to make these of mild steel and to arrange for them to be easily removed when they become scored. (Figure No. 16).

At the drawing board stages great care is always taken to ensure that the lead to and from towing bitts to the appropriate roller leads and mooring ports is perfectly straight and clear, and when the tug is fitted with a stem

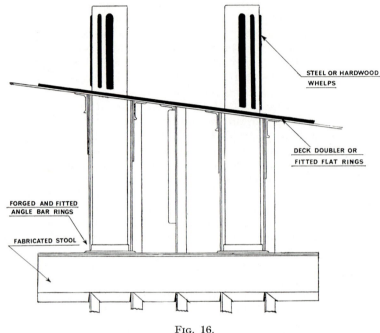

STEEL OR HARDWOOD
WHELPS

DECK DOUBLER OR
FITTED FLAT RINGS

FORGED AND FITTED
ANGLE BAR RINGS

FABRICATED STOOL

FIG. 16.
Forward Alongside Towing Bitts

head towing port and bollard it is important to see that the lead for backing up turns is fair.

It is customary to allow that all components associated with the towing bitts, such as mooring ports and fair leads, are provided at something in excess of standard. Most naval architects also provide for some stiffening of bulwarks in the way of these components.

The first requirement of an ocean-going tug, in coping with the steering function, is an aperture of sufficient size to admit the heaviest hawser, which may be envisaged for the work, through the forecastle head bulwarking. This must be arranged as closely as circumstances will allow to the stem head position and in tugs of soft stem construction this presents no difficulty whatsoever, although tugs fitted with a stem bar will be obliged to accept a somewhat less satisfactory position offset to one side of the stem bar or the other. The aperture must be suitably faired and reinforced about its circumference by means of a suitably contoured cast steel port and because of its substantial size, this aperture will require a suitable flap closure fitted on the inboard side.

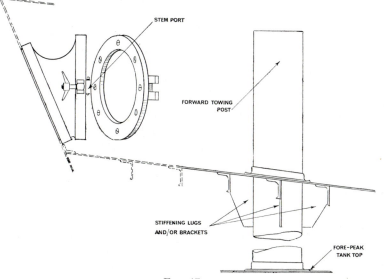

STEM PORT

FORWARD TOWING POST

STIFFENING LUGS AND/OR BRACKETS

FORE-PEAK TANK TOP

FIG. 17.

A forward towing post, of the same character and dimensions as the other fittings already described, is suitably arranged in the eyes of the tug to accept the medium, via the stem port, through which the tug will exert its steering influence. This post is passed through the deck and is heeled upon a breast hook or upon the fore-peak tank top and secured as has already been described. Some difficulty is occasionally experienced in providing a truly satisfactory position for this post owing to some congestion of appliances, and a resulting conflict for design priority, in the eyes of the vessel. Whilst it is appreciated that a proper location for the windlass is of prime importance and that cable leads, etc. demand the first place in design considerations, due allowances must be made for the installation of a stem towing post because if the post is fitted too far forward, the working of large hawsers about it becomes so difficult as to verge upon the impracticable. (Figure No. 17).

CHAPTER III

The Automatic Towing Winch

Before proceeding with a description of this important towage appliance, and its many features of interest, it is perhaps permissible to enumerate the more prominent of the many advantages which derive from its installation in modern tugs for ocean service.

1. Towing winches afford efficient and convenient stowage for steel wire tow-ropes of all practical lengths and sizes in such a manner as to render them readily available for immediate use at all times.

2. Such winches permit any desired scope of towing wire to be veered with an optimum economy of manpower, under conditions of maximum safety for the personnel involved. Furthermore, the even more laborious processes of tow rope recovery are most significantly reduced when winches are available.

3. The automatic compensating device, which is associated with most versions of the towing winch, greatly reduces the stresses of ocean towage, not only in terms of the mechanical stresses suffered by towing media and their associated impedimenta, but also those stresses of a somewhat different, but possibly even more critical variety which the vagaries of towage impose upon the responsible personnel. Because of this the whole business of ocean towage is rendered more efficient, more expeditious and immeasurably more economical in terms of man hours expended and in the preservation of useful life in towing media.

Because there is an impression prevalent among seamen that the automatic towing winch is a comparatively recent innovation . . . (and it never fails to amaze that such an impression is possibly the rule rather than the exception) . . . it must be observed that these winches have formed a part of ocean-going tug equipment for more than a half century. Early examples, many of which are still in use and rendering valuable service, were all steam powered, deriving their automatic compensatory function from an automatic steam shut off valve. So that this style of equipment shall not be lost to view

Fig. 18.

A Key to Fig. 18, accompanying the chapter on the 'Automatic Winch,'
part of Chapter II of Section 2 of the book 'Ocean Tugs.'

(a) Towing Wire Barrel.   (b) Automatic Spooling Gear.   (cd) Spooling Gear Drive
and Manual Adjustment Control.   (e) Automatic Cut-Off Valve.   (f) Manual
Friction Brake.   (g) Spragging Apertures.

Acknowledgements to Clarke, Chapman & Co., Ltd., Gateshead-on-Tyne.

whilst more modern gear is under review, the Figures 18 and 19 are included
to illustrate a steam powered winch and a typical example of a cut off valve.

A Description of the Automatic Steam Cut Off Valve
as used in association with Steam Powered Towing Winches

Figure No. 19 endeavours to provide a somewhat simplified illustration
of an automatic cut off valve assembly such as is commonly used to provide
automatic compensatory characteristics to towing winches.   The assembly

E

is ordinarily located upon the fabric of the winch at the position 'E' in the Figure 19. The operation of the device proceeds as follows.

With the winch in manual control the desired scope of towing wire is veered to the tow and towing continues thereafter, in manual control to a fixed length of towing medium, until full way is imparted to the vessel, or other floating object, which is in tow and conditions have settled. The winch is then engaged into the 'Heave In' position by means of the manual gear lever, and the drive from the Bevel Gear-Wheel 'M,' which is actuated by a similar wheel located upon the end of the barrel shaft, is engaged with

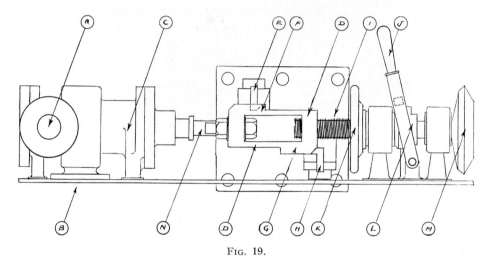

Fig. 19.

the Screwed Spindle 'I' by means of the Dog Clutch 'L' through the movement of the Clutch Lever 'J,' an alignment of the two parts of the clutch having been achieved through the Hand-wheel 'K.' The Screwed Spindle 'I' is arranged to engage into a Nut 'D' and this nut is provided with two Lugs or Projections 'F' and 'G' designed to engage with their corresponding Pawls 'E' and 'H.'

It must be noted that main steam supply to the winch feeds through the inlet 'A' and the port 'B' under both manual and automatic control, the valve 'C' being left in the full open position when the shut-off valve clutch is disengaged. When the clutch is engaged, when automatic control is required, the steam passing through the chest is thereafter controlled by the position of the valve within the chest according to the direction of the rotation of the winch barrel.

Once automatic control has been imposed, any subsequent movement of the winch barrel in the 'Paying Out' direction, as may be induced by any increase of towage stress, will cause the Screwed Spindle to rotate in a clockwise direction but the Nut 'D' will not revolve with it because the Lug 'F' will be in contact with its associated Pawl 'E.' The Nut must therefore traverse to the right carrying with it the Valve Spindle 'N' thus opening the Throttle Valve within the Steam Chest 'C.' Steam is thereafter supplied at a volume proportional to the opening of the valve thus increasing resistance to the forces causing the winch to overhaul. If such forces maintain so that the winch continues to pay out the Lug 'F' will traverse clear of its associated Pawl, thus allowing the Nut to revolve and so preventing any further lineal movement of the Valve past the normal fully open position.

When the power of the winch overcomes the stalling strain the winch barrel will commence to revolve in the 'Heave-In' direction, so imparting an anti-clockwise movement to the Screwed Spindle via the clutch and gearing. The Lug 'G' will now bear upon its associated Pawl 'H'; once more the Nut 'D' will be prevented from assuming this movement so that it, together with the Valve Stem and Valve will traverse to the left thus cutting off the supply of steam to the cylinders until the fully closed position is reached. Again any further rotation of the winch barrel due to momentum, will not be imparted to the valve because in the closed position the Lug 'G' will have passed its Pawl to allow the nut to revolve with the spindle.

Once a tow is fully settled, however, winch barrel rotation, in either direction, reduces to a very few turns except under circumstances of adverse weather, the main steam supply to the winch being adjusted to the changing demands of day to day conditions of towage.

The modern electrically powered towing winch is, however, a much more refined and complex, yet nevertheless rugged installation which is currently produced so as to afford a more sophisticated service than any of its predecessors. There are, for instance, versions of the basic winch type presently available which mount two towing wires which may, if necessary, be of different sizes and lengths. Duplex installations are now constructed where the two barrels may be either separately powered or arranged to share a common power plant. Similar dispositions apply to both the automatic compensatory devices and to the spooling gear, so that it is possible to satisfy a very wide diversity of demand. The measure of flexibility which such alternatives provide clearly reduces much of the practical difficulties which always attend upon the acceptance of multiple towage responsibilities.

For the purposes of this discussion it is proposed to exemplify a modern single barrelled model of automatic towing winch from the workshops of one of Britain's foremost specialists in this branch of marine engineering, but before entering upon this discussion the point must be made that, for all practical purposes, each towing winch is literally a 'one off' order to satisfy the conceptions of towage duty as may be professed by the individual operator in terms of dynamic duty, power and the scope and size of towing media to be mounted. The resulting variety in supply and demand find illustration in the tabulation offered in Figure No. 20.

Figure No. 20

A Tabulation of Typical Towing Winch Detail

Main Motor Horse Power	Dynamic Duty of the Winch	Render Duty	Maximum Pull when Spragged
30 H.P.	$7\frac{1}{2}$ tons at 40′ p.m.	20 tons	90 tons
40 H.P.	12 tons at 34′ p.m.	27·5 tons	90 tons
50 H.P.	15 tons at 34′ p.m.	34·0 tons	90 tons
50 H.P.	11 tons at 49′ p.m.	34 0 tons	90 tons
50 H.P.	13 tons at 38′ p.m.	None	90 tons
55 H.P.	35 tons at $15\frac{1}{2}$′ p.m.	25·0 tons	90 tons
	11 tons at 49′ p.m.		
125 H.P.	35 tons at 35′ p.m.	60·0 tons	90 tons
240 H.P.	60 tons at 41′ p.m.	60·0 tons	300 tons

The tabulation shows that certain operators attach the greatest importance to the basic loading capacity of their winches, whilst others require more consideration for the performance of the appliance in terms of the rate of tow-line recovery. These clearly indicate certain variations on spur gear ratios and in the power of the main motor. All operators demand, however, a very high standard of reliability in towing winches for use in their tugs, and the adequacy of the supply to the demand is substantially established in that there is not one single example of failure, in any of the main winch components, in the case of machinery from the workshops chosen for this discussion.

The automatic towing winch is basically a massively constructed heavy duty winch having two shafts driven from an electric motor of a capacity appropriate to the duty envisaged. One of these shafts transmits power to a central towing barrel via a train of spur gears of a ratio proportionate to the dynamic duty which has been specified by the operator. The second shaft provides power for one or two warping heads, of a diameter and width appropriate to the fibre or nylon springs associated with the class of tug concerned. The winch chassis containing these arrangements, in all of their variety, the shafting, gearing, journals and bearings are all designed, constructed and assembled to such standards, and strength, as to permit the winch to be spragged under certain conditions to withstand a maximum pull many times in excess of the duty of the winch. Besides the integrity required to cope with the extremely rigorous duties involved in ocean towage such winches are fitted with certain auxiliary features which allow of facilities which are as valuable as they are ingenious. They are:—

1. A manual and automatic switching apparatus which allows the winch to be operated orthodoxically for a manually controlled 'heave in' or 'pay out,' or to be changed over at will to another form of control so that these same functions may be applied automatically to compensate for the varying stresses imposed by towage under open seas conditions.

2. An automatic brake, which is magnetically operated, and which functions in conjunction with the manual and operating switching device.

3. A fully automatic spooling device which ensures that the towing wire will feed on to, and off from, the central barrel of the winch without over-riding or fouling.

4. An automatic centrifugal brake which prevents any critical over-speeding of the main motor.

5. A locking sprag.

6. A gear driven indicator.

A description of these facilities follows.

1. The Automatic Towing Winch Master Controller

To fully satisfy the requirements of its function a towing winch must be immediately available in orthodox winch manual control for certain aspects of that function, with the automatic compensating characteristic equally available upon demand as an alternative service. This dual facility is

FIG. 21.

Automatic Towing Winch from forward port-side.

Acknowledgements to Clarke, Chapman & Co., Ltd., Gateshead-on-Tyne.

A KEY TO FIG. NOS. 21, 22, 23 AND 24.

(*a*) Main Electric Motor. (*b*) Magnetic Brake. (*c*) Towing Wire Barrel. (*d*) Warping Head. (*e*) Change-over control for isolating the Towing Wire Barrel when Warping End only is required to be used. (*f*) Manual Friction Brake. (*g*) Automatic Control Gear Drive. (*h*) Indicator. (*i*) Automatic Centrifugal Brake. (*j*) Spooling Gear. (*k*) Spooling Gear Manual Adjustment Control. (*l*) Spooling Gear Locking Device. (*m*) Locking Sprag Housing. (*n*) Bevel Gear Drive to Automatic Control. (*o*) Manual Control and Setting Control for Automatic gear. (*p*) Change-over Switch Manual to Automatic and vice-versa. (*q*) Emergency Stop Switch.

provided through the medium of a master controller in the form of a barrel switch which is, with its various auxiliary and ancillary components, usually contained in a watertight case which may be conveniently located to one side of the winch. This barrel switch is designed so as to provide an equal number of progressive stages of control in each direction of 'heave in' and 'pay out' . . . (usually twelve to fourteen) . . . whilst the winch is in manual

Fig. 22.

Automatic Towing Winch from forward Starboard side

Acknowledgements to Clarke, Chapman & Co., Ltd., Gateshead-on-Tyne.

control. Automatic control is allowed by about one half of these control stages on the barrel switch but in the 'heave in' direction only. It must be understood that the precise number of stages, in either of the manual or automatic control applications in towing winches, varies from model to model in accordance with individual operator's conceptions of the dynamic duty of this sort of machinery in his vessels. A description of the practical arrangements devised to permit this duality of service follows herewith.

Figure No. 24 shows a typical master controller installation. With power available to this controller, either of the functions of manual or automatic control are selectable at will by the movement of the lever 'P' to one or the other of the positions marked with the self explanatory labels 'Hand' or 'Auto.' When the lever is moved to the 'Hand' position, the controller Handwheel is connected, by means of a clutch, to the barrel switch. Thereafter a clock-wise movement of the hand wheel 'O' activates the winch

Fig. 23.

Automatic Towing Winch from direction aft

Acknowledgements to Clarke, Chapman & Co., Ltd., Gateshead-on-Tyne.

through the progressive stages allowed by the barrel switch from a creep to the full 'heave in' rate according to the loading. Conversely an anti-clockwise rotation of the handwheel provides a similar progression of rate in the 'pay out' direction. The handwheel may be used for the manual control of the warping ends only, with the towing winch barrel isolated, when this is desired by the appropriate manipulation of the control 'E' shown in Figures No.'s 21 and 22.

When the automatic surge and recover function is required, the lever 'P' must be moved into the 'Auto' position but it is a most noteworthy circumstance that the lever cannot be moved to the desired position unless the Handwheel control 'O' is within the scope of 'heave in' stages allotted for Automatic control. This safeguard being achieved through an interlocking

Fig. 24.

Automatic Towing Winch Master Controller Installation.

Acknowledgements to Clarke, Chapman & Co., Ltd., Gateshead-on-Tyne.

device incorporated into the change-over mechanism. When the lever 'P' assumes the 'Auto' position, the handwheel 'O' is disconnected from the barrel switch and all subsequent movement originates from the rotation of the central towing winch barrel 'C' in Figures 21 and 22.

This rotation is initially transfered to the master controller by means of the bevel gear 'G' mounted directly upon the end of the towing winch barrel shaft, (Figures 21 and 24). This wheel engages with its Mate, bevel gear 'N,' mounted upon the exterior casing of the master controller to drive a reduction gear which transmits the adjusted movement to an epicyclic unit which provides the final drive to the barrel switch. The actual mechanical connection between the automatic drive and the barrel assembly is achieved by means of a lock which is contrived by the utilisation of the epicyclic gear housing in the role of a brake sheave to be gripped securely by a pair of caliper action brake shoes when the selector lever is moved into the 'Auto' position. Once the automatic gear is locked on, any subsequent rotation of the central towing barrel is transmitted proportionally to the barrel switch so that if the winch pays out towing gear, the rotation so induced advances the switch barrel so as to increase tension until a balance is reached and, of course, vice versa.

When the winch is to be operated automatically, the practical procedure is simplicity itself. With power to the main motor in supply, the lever 'P' is put to the 'Hand' control position and any desired scope of towing medium is veered to the tow; when the vessel or object towed has settled to the scope allowed, the Hand-wheel control is eased back until the winch is rendering slightly under the stress of towage. The Hand-wheel is then advanced one stage only and the lever 'P' is put into the 'Auto' position. Any subsequent movement of the winch thereafter is translated into appropriate compensatory movements of the barrel switch. Any adjustment to towing scope which may be required after the initial setting is made by returning the selector lever to 'Hand,' placing in hand the desired adjustment and thereafter returning the Hand-wheel control to the appropriate stage of automatic control before returning the selector lever to the 'Auto' position. Should the stress of towing mount higher than the highest automatic stage, a condition which ordinarily only obtains under conditions of crisis, or severely adverse weather, the automatic gear is arranged to cut out and devices are available to relate this condition to a warning light or sound signal.

Main electric motors for towage winch employment are conventional direct current shunt wound machines of a power and capacity appropriate to the dynamic duty specified by the operator. (Figures No. 21 and 22 'A').

In view of their conditions of service they are totally enclosed and water-tight. The control exerted from the Master Controller acts via Ward Leonard principles to strengthen or weaken a generator motor shunt field to provide voltage control for the main motor; reversal being obtained by changing the polarity of the shunt field. All towage winch circuits are provided with an Emergency stop switch, (Figure No. 24, 'Q'), which, when operated, shuts the whole system down forthwith.

2. The Automatic Brake

Towing winches, like all other winch types, and similar machinery designed to endure stress and loading under varying conditions of service, must be provided with the facility of sustaining load or stress in the event of a current failure, or with main motors at rest. This facility is afforded, in the case of the winches under review, by means of an automatic magnetic brake which is arranged integrally with the Main Motor. (Figures No. 21 and 22 'B').

The working principle of this type of brake, in the broadest possible terms, is that as long as full mains current is available at the motor, powerful electro-magnets are energised so as to hold plates bearing frictional material away from, and clear of, a brake disc securely attached to the motor shaft. Should current be removed from the winch motor, by accident or design, these electro-magnets de-energise to allow a set of powerful springs to exert a sufficiency of power to apply the plates bearing the frictional material against the brake disc with such force as to prevent any further rotation of the motor, and thus the winch, until the brake is either released mechanically or power is restored to re-energise the electro-magnets and thus release the brake disc. The value of such an accessory needs no emphasis in matter descriptive of winches for ocean towage duties.

3. The Automatic Spooling Device

It is entirely self-evident that towing wires must be stowed upon towing winch barrels in tight precise turns so that the hauling part cannot, through the stress of towage, be ripped down through the supporting turns on the barrel to inflict lacerations upon these turns or to set up even more complicated stresses. It is equally self-evident that the towing wire must never be veered off from the barrel in any plane other than that of right angles to the horizontal axis of the barrel, if similar damage is to be averted.

The first contingency is provided for in the design of the barrel proper. Each winch barrel is precisely built to dimensions calculated so as to exactly contain a carefully related number of turns of wire, as specified by the operator, so that the last turn on the barrel lays fairly up to the barrel flange so as to feed back across the barrel with the second layer of turns bearing exactly, turn for turn, upon the one beneath. Where no departure from the original size of wire specified for use with a particular winch is ever intended, winch barrels are sometimes provided with a helical groove to accept the first layer of wire to great advantage.

The second consideration is provided for with the utilisation of Automatic Spooling Gear. (Figure No. 23 'J'). This device locates upon the aftward side of the winch and consists of a massive box fairlead structure comprising a strong steel carriage bearing three large roller members, two vertical and one horizontal, in a 'U' formation. This fairlead is arranged to traverse, in a horizontal plane backwards and forwards, across the after face of the winch barrel upon a pair of substantial guide rails, movement being imparted by a driving shaft with two spiral grooves, one left-handed and the other right, deeply and accurately incised into its surface, mounted midway between the guide rails. This shaft is rotated through the medium of spur and bevel gearing from the *barrel shaft*, (Figure No. 23 'K'), the backward and forward motion being imparted to the fairlead when a stout key, arranged upon the underside of that unit is inserted into the grooving by means of the Handwheel 'L' in Figure 23. The gear ratio incorporated into this drive is calculated so as to advance the fairlead carriage across the face of the barrel a distance exactly equal to the diameter of the towing wire for each complete revolution of the barrel. This drive is also provided with a clutch and a handwheel for setting up and adjusting the alignment of the fairlead.

From the foregoing it will be apparent that should changing circumstances of service require a different size of wire for any particular winch, it may be necessary to change the barrel, it will, however, always be necessary to modify the gear ratio in the spooling gear drive   The lead of the towing wire from the majority of towing winches in use today is arranged from the upper side of the barrel so that the spooling gear traverses on the upper sides of the guide rails. When circumstances demand the installation of a winch at a higher level in relation to the other associated components, the spooling gear may be slung under the guide rails to allow of a low-level haul-off to some advantage. The only modification enjoined by this procedure consists of adding a keep to the vertical rollers for the retention of the wire when slack. (Figure No. 18).

4. The Automatic Centrifugal Brake

Salvage and Ocean Towage Operations are occasionally progressed under circumstances of considerable hazard; upon occasion vessels, or other floating objects under towage, founder at very short notice in water of a depth considerably in excess of the length of the towing medium so that a grave risk to the tug results. Reflotation exercises also occasionally place the tug or tugs involved in circumstances of danger. It is therefore not an unusual circumstance for a tug to be obliged to separate herself from the object of her attentions in some haste. Where towage is conducted from a hook or a bollard using traditional media this eventuality presents no difficulty but when a towing wire from a winch provides the connection, then a safe separation is a matter deserving of special attention.

With a towing winch in use it is manifestly quite irresponsible to isolate the central barrel and to allow the wire to take charge under the stresses which may obtain. Even if the equipment should survive the velocities which could well result, the danger and destruction which could result from a fouling or riding turn upon the barrel, or the flailing end, or ends, of a substantial wire can only be imagined. It is also clearly beyond the bounds of reason to expect personnel to endeavour some control of the situation through the use of the manually operated band friction brake under these conditions. (Figures No.'s 21 and 22 'F'). It is apparent also that any attempt to veer wire off the winch under power, whilst enduring the high stress to be expected, must be inevitably accompanied by serious damage to the main motor.

The manufacturer's solution to this difficulty consists of devising an arrangement whereby the maximum revolutions of the main motor may be restricted to a certain pre-determined limit through the employment of an automatic centrifugal brake. This device which attaches to an extension to the main winch motor shaft (Figure No. 22 'I') consists of a revolving unit which is designed to contain four spring loaded elements each arranged to contain a frictional lining. This unit revolves within a robust brake drum assembly which is effectively integrated into the winch chassis. At revolutions to be associated with normal winch working, these spring elements are of sufficient strength to hold the brake linings clear of the drum, but if some critical condition accelerates the motor beyond the limit imposed then the centrifugal force which is engendered thereby serves to apply the brake linings to the drum at a pressure proportionate to the speed imposed. This arrangement permits a tug to steam away from a sinking, or otherwise hazarded tow running the wire off from the winch barrel under the braking effect imposed by the centrifugal device augmented by the winch gearing, without ill-effect to the motor. It is, of course, apparent that the circumstances obliging the extrication movement could still overtake this expedient, but the ensuing damage would not involve the prime mover for the winch.

5. The Locking Sprag

Circumstances can, and do, arise whilst ocean towing is in progress so that a combination of the weight of the tow with adverse weather together provide stress which is in excess of both the capacity of the compensating gear or of the friction brake. The only recourse then open to the Tug Commanding Officer is to sprag his towing winch. To provide this facility the large spur wheel on the winch barrel shaft is provided with a number of evenly spaced elongated oval apertures arranged just within its periphery. Arranged to be engaged into any one of these apertures is a massive steel bolt of a section appropriate to the aperture. This bolt is mounted so as to slide within a housing of dimensions appropriate to the duty and it is inserted and withdrawn as occasion demands by means of a lever provided with a locking pin. (Figures No.'s 21, 22 and 23 'S' and 'M').

When the sprag is in position, the power may be removed from the winch because all subsequent stress is transmitted via the sprag and the barrel shaft to the winch chassis and thence to the major construction components with which the winch foundation is integrated. When the winch is spragged it is customary to withdraw the Spooling Gear key from the drive shaft.

6. Indicator

It is the current practice to provide all towing winches with an indicator (Figure No. 21 'H') which will show the amount of wire which has been veered from the barrel. Such indicators are driven from the barrel shaft through gearing of an appropriate ratio to the dimensions of the barrel and the size of the wire. The indicator detail will, of course, be inscribed according to the nationality of the tug to be fitted.

CHAPTER IV.

Fittings and Appliances Peculiar to the Ocean-Going Tug Type. Gobbing or Gogging Eyes and Bollards, Molgoggers, Towing Arches or Horses, Bulwarking Arrangements and Protective Belting Devices.

To assist in the reduction of chafe to towing media whilst carrying out ocean towage operations and to retain full control of the towing gear whilst towing, particularly whilst towing at short stay, it is customary to rig a length of substantial cordage from an afterly location in the tug, over and about the towing medium, and back inboard of the tug in such a fashion as to maintain control, either by turning the cordage up, or by taking it to a winch or capstan. This piece of cordage, which may be of steel wire, fibre rope or nylon rope, depending upon the towing medium in use, is known variously as a 'bridle,' a 'gog-rope' or a 'gob-rope.'

Convenience and efficiency demand that proper provisions be allowed for this important aspect of towage drill and naval architects generally furnish two appliances for its execution viz.:—a 'bridling, gogging or gobbing bollard' and a 'bridling, gogging or gobbing eye or lead.' These two fittings are welded to the deck, sometimes in association with a common doubler plate, as far aft in the tug as will allow of an approach from all angles, the bollard being usually arranged a few feet forward of the lead.

The bollard, which is invariably of the twin-legged cruciform pattern, may be constructed of seamless steel tubing or may be a single steel casting. The lead, or eye, is usually a steel casting devised to present two or three well faired apertures suitably to the bollard. Bearing in mind the possibility that ocean tugs may, from time to time, be obliged to tow more than one object simultaneously, the provision of three apertures in the lead will certainly be a great future convenience to tug personnel. (Figure No. 25).

Whilst ropes made from synthetic filaments, with their vast strength and minimal bulk, are becoming more and more the vogue in ocean work, their bulk still occasions practical considerations, moreover certain aspects of ocean towage still derive benefit from the bulk and weight of manila, or other vegetable fibre, towing springs and this large cordage poses a number of

67

complications to the tug personnel, not the least of which is its recovery after a towage operation is completed.

The obvious first essential in this aspect of towage operations is the provision of deck machinery adequate to the function and this may be provided by fitting a warping drum of suitable dimensions to a towing winch or by the installation of a capstan of such dimensions as to accept a sufficient number of turns of towing medium to heave directly, or lastly, a capstan of

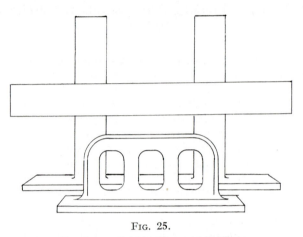

Fig. 25.

Bridling, Gogging or Gobbing Eye and Bollard.

more modest dimensions which may be used to fleet the tow ropes inboard in conjunction with a stopper and messenger. Regardless, however, of the appliances available for the actual heaving in processes, the operation cannot proceed with safety and efficiency unless some sort of fairlead can be arranged to provide a correct and constant lead to the feeding side of the capstan or winch drum from overside. Owing to the fundamental requirement for an unimpeded arc of travel for the towing media over a perfectly smooth surface all about the after parts of a tug, the provision of a suitable fairlead superior to taffrail level is clearly quite impracticable so that portable or folding devices are necessitated.

In the smaller classes of tug these 'Molgoggers,' as such portable or folding devices are named, take two forms. The first takes the shape of a single roller mounted upon a stanchion of the desired height, arranged to hinge at deck level and secure to a bulwark stay. One of these devices is arranged upon each quarter of the tug and when the towing medium is to be hove inboard the weather, or otherwise most convenient, molgogger is erected to serve as a restraining or localising lead to give a constant lead toward the winch or capstan. The second version of this appliance consists of a 'U'-shaped arrangement of three steel rollers mounted upon a robust stem designed to ship into suitable sockets located in, or just inboard of, the taffrail. This latter appliance providing the more positive lead control of the two so that enlargements and elaborations of it provide standard equipment in the heavier classes of tug.

In these larger classes of tug, with their heavier and bulkier towing media, the molgoggers are installed permanently and robustly in pairs, one at each quarter, as fitted features. The problem of keeping the after sections of tug unimpeded by obstruction being solved by arranging the triple roller heads to fold downwards and to rotate about the vertical axis, this facility being assisted by the provision of plain metal bearings at the pivotal points and counter-poise weights upon the lower sections of the assembly. Strength and stiffness to resist the not inconsiderable stresses involved are provided by mounting the rollers upon very sturdy pins within the most substantial containing material whilst the fixed parts of the assembly are most adequately bracketted to the deck in both the fore and aft and the athwartships directions. (Figure No. 26).

Earlier chapters in this work have touched upon the weight and stress endured by the towing medium and the associated appliances and of the design considerations thereby imposed upon the point of tow and the tugs' after sections. Mention has also been made concerning the establishment of the point of tow and the fact that a not inconsequential length of towing medium must always intrude inboard. All of this points up the need for making such arrangements as will prevent the towing medium from inconveniencing tug personnel in going about their duties in the towing deck, and will avert the likelihood of the inboard sections of the towing gear fouling deck erections and appliances whilst it is in movement as it accepts the variety of loading imposed by towage whilst both tug and tow make their separate and different reactions to the conditions of wind, sea and swell. The tug feature developed for this purpose is known as a 'towing horse' or 'towing arch.'

This feature is fitted into tugs in such numbers, and at such a height, as will prevent an unstressed tow-rope from fouling dominant deck erections and fittings. In the case of the larger classes of tug this will involve three or

F

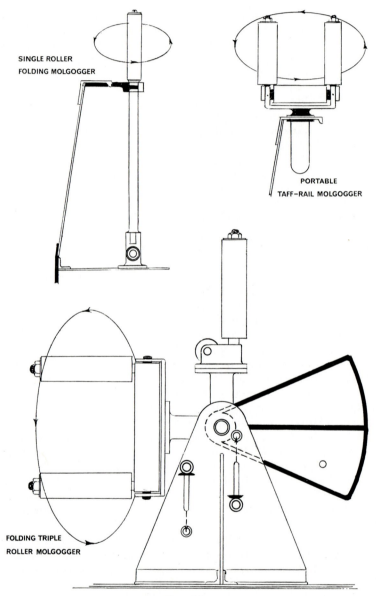

SINGLE ROLLER
FOLDING MOLGOGGER

PORTABLE
TAFF-RAIL MOLGOGGER

FOLDING TRIPLE
ROLLER MOLGOGGER

FIG. 26.
Molgogger Detail.

four arches fitted at intervals of about twenty feet. The arches are fitted, having the necessary curvature to provide the desired height, in a fair and fine arc from bulwark to bulwark. They are set up, in relation to one another, and to the taffrail, and the point of tow, so as to offer the fairest possible lead for the towing medium. In view of the very arduous nature of their function, towing arches must be of strong construction and rigidly supported. The workmanship involved in forming the bearing areas must be of a high standard so as not to produce friction provocative joints, whilst the detail of the connection between the upper surfaces of the horses and the bulwark rails must receive close and expert attention. (Figure No. 27).

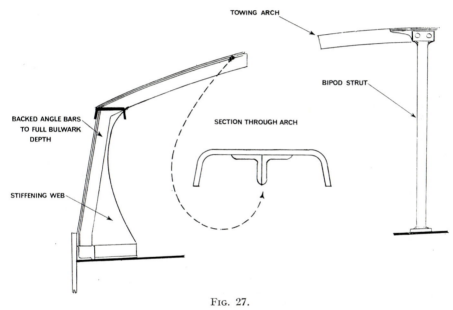

FIG. 27.

Towing Arch Detail

Figure No. 28 (a), (b), and (c) represents endeavours made towards achieving a satisfactory bulwark and belting assembly which will stand up to the rigours of tug requirements and still offer an adequacy of protection to personnel working about the decks, whilst inflicting the least amount of damage to towing media.

Bulwarks are usually fitted in ocean-going tugs to a height of at least 3 ft. 0 ins. but this height is not constant for the whole run of after deck bulwarking because most naval architects arranged a run down in height towards aft in order to ameliorate chafe effects. Bulwark plating is usually given a tumble home approximating to 3 ins. per foot of height in order to reduce vulnerability during alongside towage operations. Bulwark stanchions obtain in a variety of designs three of which are illustrated in Figure 19, these are usually provided at every third beam. It is customary to fit bulwarking at rule thickness everywhere except towards the after sections

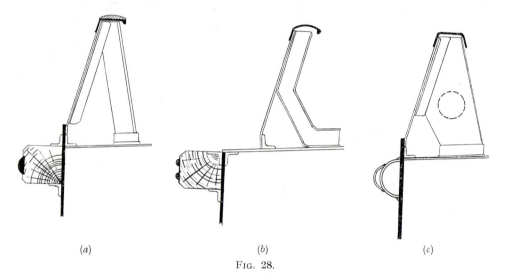

(a)   (b)   (c)

Fig. 28.

where it is usually progressively increased. Doubler plates to the full depth of the bulwarks are invariably provided in way of mooring pipes and leads. Ocean-going tugs tend to be 'half tide rocks' aft so that a generous allowance of freeing port area is the general rule, one eminent designer recommends an increase of 20 per cent, in this respect, over that required by regulation. The modern tendency is to provide freeing area in terms of long slots arranged close to the deck with the upper edges of the slot opening flanged to provide fore and aft stiffness.

This chapter ends with a note or two upon belting for ocean-going tugs. The type to be fitted to any particular tug will reflect, to a great extent, the

operator's conception of the function of his vessel. An operator who anticipates that his tug should cope with any operation that falls to its lot will
provide heavy duty belting all the way around his vessel whilst another, who
wishes only to be involved in deep-sea towage in salvage or by contract may
not call for anything other than the addition of a belt of half-round or
elliptically sectioned metal, a few inches deep, about the vessel at the level
of the main deck.

The traditional approach to belting has hitherto been to secure hard wood
about the after sections at the main deck level, and about the fore-parts of
the ship at the forecastle head level, with this latter part being extended for
the full length of the superstructure so as to provide two strakes of protection
over the mid sections. The timbering was secured to the hull by means of
through bolts or spikes to a pair of continuous angle bars secured, parallel to
one another, over the required areas by means of welding. The timber being
embedded into pitch, red lead or any one of a number of patented anti-
corrosive mixtures available, between the angle bars.

This is a tremendously expensive business at the building stage and
remains very expensive in terms of repair and maintenance throughout the
life of the tug, a circumstance which cannot improve in view of the increasing
scarcity of American Elm and other suitable timbers, in the appropriate
dimensions; and the increasing cost of labour.

Because of this, the modern tendency is to replace the timber fendering
by either a semi-circular welded steel section, welded directly to the hull, or
by rubber or plastic material of appropriate shape and dimensions fitted
between welded angle bars.

Regardless, however, of the style of belting, or the material used, it is
vitally important to provide boat skids over, and under, the belting both in
way of the davits and in vicinities where it is convenient to bring boats
alongside. Boat work in the open seas is always a very hazardous business
and the risk of capsizing boats by fouling rubbing bands, or belting, both
whilst launching and whilst drawing alongside is great, particularly when
the tug is rolling and pitching in a seaway.

CHAPTER V.

The Stability of Ocean-Going Tugs.

Stability in tugs of all types has been exhaustively discussed and described in various works devoted to the naval architecture of specialist ship types, and in technical papers read before the appropriate professional societies in the United Kingdom, Northern Europe and in the United States of America. No submission offered in this book can, therefore, pretend to be any more than the briefest possible note upon this subject, from the viewpoint of the practical tug seaman.

Because of the vertical heeling moment which can exert through the towing medium, the ocean-going tug type requires an excess of stability over that which is required in the case of other vessels of a comparable size operating under the same practical conditions. Service Constructors, and their civilian colleagues in most maritime countries, appear to be agreed that the ocean-going tug should be designed with at least 12 ins. or 30/31 cms. of metacentric height whilst in the light condition. Whilst most tug seamen would fully concur with this conclusion, seeing that a seakindly and sea-worthy vessel is thereby ensured, they are wholly convinced that a satis-factory range of stability is of paramount importance in vessels obliged to perform vigorous manoeuvres under circumstances of adverse weather. It is a thought, often expressed when ocean towing men foregather, that the special problems arising from the vigorous application of helm and engines to vessels of modest size and lively habit, whilst operating in swell and sea, have not been as exhaustively examined as they could be.

The problems associated with free surface effect upon ships' stability are specially applicable to tugs in view of the duration of towage operations resulting from a commonly low speed of advance. Ample endurance is an extremely desirable operational characteristic for all ocean tugs and the excess of tankage thereby necessitated should not, under any circumstances, be allowed to constitute a potential operational hazard for want of sufficient and efficient internal subdivision. Informed design should also provide such arrangement of pumps and pipelines as will give the fullest flexibility in the use of fuel, fuel/ballast and fresh water tankage so as to ensure that all compartments may be either quite empty, or pressed full, as far as may be practicable, according to the conditions obtaining.

A point, possibly worthy of mention, concerns the relatively fine ends of ocean tugs. Because of this, the peak tanks tend to provide something of a mixed blessing in terms of ballastage. Their triangular section occasions a very large free surface relative to their capacity when they are not pressed full; furthermore because of this circumstance the centres of gravity of filled peak tanks is so high, in relation to the C. of G. of the tug as a whole, that the product of their filling may be an impairment of metacentric height rather than an improvement, especially with the tug in the light condition. It follows, therefore, that any improvement to towage performance sought by any increase in displacement tonnage should not derive from the implementation of peak tanks before the effect of such action has been considered in relation to the stability of the tug.

THE OCEAN-GOING TUG AND THE PRACTICE OF OCEAN RESCUE.

SECTION 3.

CHAPTER I.

The Towing Medium.

Towing mediums comprise varying lengths and differing combinations, depending upon practical considerations, of ropes manufactured to contemporary finest standards from either natural vegetable fibres, steel wires or filaments of man-made material. A description of these now follows, commencing with tow ropes made from vegetable fibre.

1. Fibre Towing Springs.

The supreme requirement in fibre springs must necessarily be great strength to withstand the heavy stresses which must be anticpiated when large vessels, and other floating objects, are in process of being towed through the varying conditions of weather experienced in the open seas. Secondary, but nevertheless important qualities desired are a high degree of elasticity . . . (of which more later) . . . and the property of resistance to the effects of long exposure to possible climatic extremes and salt water.

The strength of ropes is principally determined by the tensile strength of the raw fibres which, when processed and spun together, provide the rope's yarns. The continuity of strength in these yarns being necessarily dependent upon the uniformity, in both length and strength, of these raw fibres. It follows then that the ideal vegetable fibre for tow rope construction must needs be both strong and long. Manila fibre has been found to be the most suitable fibre for the purpose because of its high tensile strength and

76

because the average length of the raw fibres is about eight feet. In comparison the equivalent figures for sisal fibre, the next best, are about 80 per cent in terms of tensile strength with an average fibre length of four to five feet.

Concerning the property of elasticity; there is, for all practical purposes, no elasticity in any of the vegetable rope-making fibres, so that any elastic properties found in this type of rope must be induced by the various processes used in its manufacture.

The series of operations which are carried out in order to convert a mass of vegetable fibres into rope are mainly processes of a rotary nature. Thus, the fibres, after cleaning and combing, are spun together to make yarns, these yarns are in turn rotated together to form them up into rope strands; the strands thus formed being once more rotated in order to impart the desired degree of twist into them before they can be laid up together to form rope. These rotationary processes are alternately right and left-handed, finishing up right-handed in right-hand laid rope in order to achieve a balanced dissemination of stresses within the rope whilst working.

The primary object of all of this twist or torque in all of the component parts of the rope is to induce a high degree of friction which provides the 'sticking power' which literally holds the rope together. A secondary effect however, attends upon all of this rotational treatment in that a finished hawser-laid rope is only about 70 per cent of the length of the strands from which it is formed. This loss represents a measure of extensibility which may be drawn upon when the rope is under stress. The use of the word 'extensibility' must be especially noted, because the expressions 'elasticity' and 'extensibility' are not synonomous.

Whilst a manila fibre rope is only extensible in the immediate sense, it does possess limited elastic properties in that it will recover to a considerable degree after extension if it is left to rest. It will also recover immediately from extensions of small magnitude and of short duration.

Because it was recognised and respected that it was only the laying up processes which induced elasticity into fibre ropes, the principle was carried one step further in the construction of special cordage for the use of ocean-going tugs, in an attempt to provide an enhanced measure of this desirable property to meet their special needs.   This special cordage became known as 'cable laid' rope and consists of three hawser laid ropes laid up together to form one large rope.   In rope of this construction the property of extensibility is considerably increased over hawser laid rope of the same dimensions, because the finished rope is only a little over one half of the length of the strands making up the individual ropes within it.   This extensibility gained must, however, be balanced against a certain loss of strength, when compared with a hawser laid rope of the same size; the loss amounts to about 40 per cent, which has always been accepted by the calling in view of the extensibility gained.

It is the common practice for the three ropes used in making cable laid rope to be specially laid up left-handed in order that the finished rope shall be an orthodox right-hand lay.

The flexibility of fibre ropes lies partly in the character of the raw fibre and partly as a result of manufacturing processes.   Referring to the former, one has only to handle best manila side by side with best sisal to see what is meant, but the manufacturing processes consist of varying the degree of twist applied to produce hard or soft laid rope, the soft laid having received less twist to produce a more flexible rope.

Cable laid manila and sisal towing springs are made in various sizes and lengths to suit the needs of the different classes and types of tug.   The usual maximum length is one hundred and twenty fathoms but circumferences rise in 1 inch stages from 10 ins. to 26 ins.; all springs can be obtained ready fitted with a thimble and link spliced into each end, or with a leather sleeved soft eye in one end and a thimble or thimble and link at the other.

2.   Steel Wire Towing Ropes.

Steel wire tow ropes must clearly possess the same qualities of strength, durability and ductility demanded of their fibre counterparts.

The strength of the rope is primarily vested in the chemical composition of the cast steel billet from which the individual wires which form the strands of the rope are drawn.   The quantity and quality of the different ingredients

which combine to form this billet and the processes involved in its manufacture are all part of the industrial expertise of the manufacturer as are,
of course, the involved processes required to roll and draw the billet into a
series of apparently endless wires all of precisely uniform diameter.

The flexibility of steel wire ropes for towing is proportional to the number
of wires employed to build up each strand. Flexibility in the larger sizes of
ropes is maintained by increasing this number. Were this not done the
larger ropes would become progressively stiffer, in direct ratio with their
size, by virtue of the greater diameter of the individual wires so as to make
them practically unhandleable. Wire ropes for towage work of up to 5 ins.
circumference are usually of 6×37 construction whilst those of over this
size are of 6×61. They are supplied from 3 ins. circumference upwards in
increments of $\frac{1}{2}$ in. to 5 ins. and thereafter to order in increments of $\frac{1}{4}$ in.
The manufacturers make up tow ropes to any required length and fit them
with thimbled eyes at one or both ends. Towing winch wires, in lengths of
from 250 fathoms to 450 fathoms are supplied with a thimbled eye in one
end with the other whipped, tapered or ferruled to order.

Because of the practical conditions attending upon the use, and stowage,
of steel wire ropes for towing, all expert opinion is entirely unanimous in the
demand that such ropes shall be protected by one of the more superior and
specialised galvanising processes. The protective material used should be
zinc, at the highest practicable level of purity, deposited electrolytically
upon the wire at a uniform concentricity to a substantial thickness, so that
the ductility of the coating, and its adhesion to the wire, is consistent with
towage requirements.

The zinc coating provided by this process provides a sacrificial protection
to the underlying wire, both in terms of corrosion and abrasion, from which
it is evident that the heavier the deposit of zinc, the longer the period of
protection.

3. Tow Ropes Manufactured from Man-Made Fibres.

The term *synthetic* immediately calls to mind the various man-made
substances which have been evolved from time to time to take the place of

natural materials which have become scarce or inaccessible for any one of a variety of reasons. In the greater majority of cases, such synthetic materials are regarded only as inferior substitutes and to be discarded the moment that the true material becomes available once more. In the sphere of rope manufacture however, this was, most definitely, not the case because the synthetic materials evolved proved to be not only satisfactory but to be immeasurably superior to the very finest of natural rope-making fibres hitherto used.

In the first part of this Chapter the qualities of great strength, flexibility, elasticity and durability were all quoted as being essential in fibre ropes and manila fibre was instanced as being the fibre possessing these qualities in the highest degree. It was pointed out that there was no inherent elasticity in any vegetable fibre and that a basic weakness of all fibre ropes lay in the fact that the yarns were made up of relatively short lengths of fibre, further, that the strength of ropes depended to a large extent upon uniformity in length and strength of the individual fibres.

From this it is evident that the ideal towing medium would be made up of filaments . . . or fibres . . . call them what you will, which were of an absolutely constant diameter throughout their length, such length being sufficient to make a rope to any desired length without any joins whatsoever. If to this is added the proviso that these filaments should be immensely strong, truly elastic, extremely flexible and proof against decay, then we possess the ideal rope making medium . . . what we have in fact is Nylon.

In a work of this description it is not necessary to describe the manufacturing processes involved in the production of nylon filaments for rope construction and any reader who requires this information is referred to the appropriate text books, but it is quite difficult to write about these man-made ropes without enthusiasm and, of course, bias. The following general observations are however offered:—

Nylon rope is immune to decay and rot, it is more than twice as strong as fibre rope of the same dimensions. It is extensible in the filament, as well as in construction, and has vastly superior elastic and recuperative qualities to equivalent manila, tow ropes can therefore be hawser-laid to no disadvantage. Nylon is not attacked by mould or bacteria so that it can be stowed away wet and it does not absorb any appreciable amount of water when immersed so that it does not swell nor does it suffer in low temperature operating conditions.

4. Tow Rope Accessories.

In connection with any description of towing media, it is perhaps proper to give some description of the fittings which are incorporated into towing gear . . . Viz. the thimbles, thimble and link sets and shackles, which are required to form proper connections between the separate components of a tow rope and to the terminal points of attachment to the tug and object towed respectively.

Thimbles.

Whenever any eye-splice formed in rope is required to endure stress or frictional wear, it is customary to protect the area by introducing a thimble. In the ordinary run of maritime requirement, thimbles are mostly appreciated as a means of reducing risk of failure associated with frictional wear, but in the field of ocean towage, as in the very heavy lift business, the humble thimble has a much more vital function. Here the thimble is required to literally reinforce the splice to its correct shape against stresses which are striving to close it in flat. The importance of properly holding a splice to its correct shape cannot be too highly emphasised for, should a thimble collapse, the loss of a tow might well follow.

When supported to its correct shape and proper radius by a properly proportioned and well fitted thimble, the crown of an eye-splice is well able to endure the stresses of towing up to the ultimate strength of the splice, but if the thimble should collapse, the crown of the eye would be obliged to support the loading about the considerably reduced diameter offered by the towing hook itself, the shackle pin or the link depending upon the location of the casualty. Such diminishment could, under the stresses of towage, prove so marked as to induce a progressive breakdown of the filaments in this locality.

A secondary hazard of course arises from lacerations which fragments of broken thimble may inflict upon adjacent sections of rope.

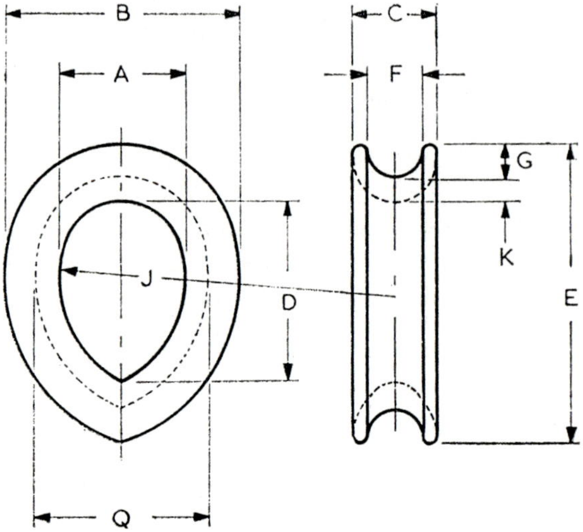

FIG. 29(*a*). Ordinary Thimble.

Nominal size (Dia. of rope)	For ropes sized by circ.	A	B	C	D	E	F	G	J (approx.)	K	Q
in.	in.	in.	in.	in. *(min.)	in.	in.	in. *(min.)	in.	in.	in. *(min.)	in.
$\frac{15}{16}$	3	$2\frac{1}{2}$	$4\frac{5}{16}$	$1\frac{5}{16}$	$3\frac{5}{8}$	$5\frac{3}{4}$	1	$\frac{1}{2}$	7	$\frac{13}{32}$	$\frac{5}{16}$
1	$3\frac{1}{4}$	$2\frac{3}{4}$	$4\frac{11}{16}$	$1\frac{3}{8}$	$4\frac{1}{4}$	$6\frac{3}{8}$	$1\frac{1}{16}$	$\frac{9}{16}$	8	$\frac{13}{32}$	$3\frac{9}{16}$
$1\frac{1}{8}$	$3\frac{1}{2}$	3	$5\frac{1}{4}$	$1\frac{1}{2}$	$4\frac{3}{8}$	7	$1\frac{1}{8}$	$\frac{5}{8}$	9	$\frac{1}{2}$	4
$1\frac{1}{4}$	4	$3\frac{3}{4}$	6	$1\frac{5}{8}$	$5\frac{1}{4}$	$7\frac{3}{4}$	$1\frac{5}{16}$	$\frac{5}{8}$	10	$\frac{1}{2}$	$4\frac{3}{4}$
$1\frac{3}{8}$	$4\frac{1}{2}$	$4\frac{1}{8}$	$6\frac{7}{8}$	$1\frac{7}{8}$	6	9	$1\frac{1}{2}$	$\frac{3}{4}$	12	$\frac{5}{8}$	$5\frac{3}{8}$
$1\frac{1}{2}$	5	$4\frac{1}{2}$	$7\frac{3}{4}$	$2\frac{1}{8}$	$6\frac{1}{2}$	10	$1\frac{5}{8}$	$\frac{15}{16}$	13	$\frac{11}{16}$	$5\frac{7}{8}$
$1\frac{5}{8}$	$5\frac{1}{4}$	$4\frac{1}{2}$	$7\frac{3}{4}$	$2\frac{3}{16}$	$6\frac{1}{2}$	10	$1\frac{11}{16}$	$\frac{15}{16}$	13	$\frac{11}{16}$	$5\frac{7}{8}$
				—			—			—	
$1\frac{3}{4}$	$5\frac{1}{2}$	5	9	$2\frac{1}{4}$	7	$11\frac{1}{4}$	2	1	14	1	7
$1\frac{7}{8}$	6	$5\frac{1}{4}$	$9\frac{3}{4}$	$2\frac{5}{8}$	$7\frac{1}{2}$	$12\frac{1}{2}$	$2\frac{3}{8}$	$1\frac{1}{8}$	15	$1\frac{1}{8}$	$7\frac{1}{2}$
2	$6\frac{1}{4}$	$5\frac{1}{2}$	$10\frac{1}{8}$	$2\frac{3}{4}$	8	13	$2\frac{1}{2}$	$1\frac{3}{16}$	16	$1\frac{1}{8}$	$7\frac{3}{4}$
$2\frac{1}{8}$	$6\frac{1}{2}$	$5\frac{1}{2}$	$10\frac{1}{8}$	$2\frac{3}{4}$	8	13	$2\frac{1}{2}$	$1\frac{3}{16}$	16	$1\frac{1}{8}$	$7\frac{3}{4}$
$2\frac{1}{4}$	7	$5\frac{3}{4}$	$10\frac{5}{8}$	3	$8\frac{1}{2}$	14	$2\frac{5}{8}$	$1\frac{1}{4}$	17	$1\frac{3}{16}$	$8\frac{1}{8}$
$2\frac{1}{2}$	8	$6\frac{1}{4}$	$12\frac{1}{4}$	$3\frac{3}{4}$	$9\frac{1}{2}$	$16\frac{1}{4}$	$2\frac{3}{4}$	$1\frac{3}{4}$	18	$1\frac{1}{4}$	$8\frac{3}{4}$
$2\frac{3}{4}$	9	8	16	$4\frac{3}{4}$	$10\frac{3}{4}$	$19\frac{3}{4}$	$3\frac{3}{8}$	$2\frac{3}{8}$	19	$1\frac{5}{8}$	$11\frac{1}{4}$

The Dimensions of the Thimble and Link Sets used in Fibre Springs

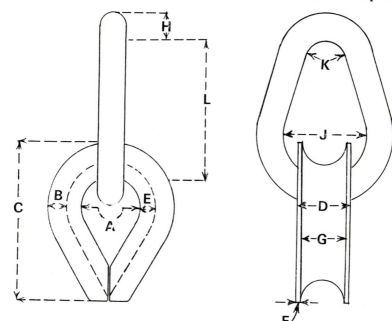

FIG. 29(b).   Thimble and Link Set.

Circ. of Rope "	A. "	B. "	C. "	D. "	E. "	F. "	G. "	H. "	J. "	K. "	L. "
10″	4	$1\frac{9}{16}$	11	$3\frac{1}{2}$	$\frac{7}{8}$	$\frac{1}{4}$	4	2	$6\frac{1}{8}$	$3\frac{5}{8}$	$10\frac{1}{8}$
12″	$4\frac{3}{4}$	$1\frac{13}{16}$	$12\frac{7}{8}$	$4\frac{1}{4}$	1	$\frac{1}{4}$	$4\frac{3}{4}$	$2\frac{3}{8}$	7	$4\frac{1}{8}$	$11\frac{7}{8}$
14″	$5\frac{1}{2}$	$2\frac{1}{8}$	$14\frac{7}{8}$	5	$1\frac{1}{8}$	$\frac{5}{16}$	$5\frac{5}{8}$	$2\frac{5}{8}$	8	8	$13\frac{1}{8}$
16″	$6\frac{1}{4}$	$2\frac{3}{8}$	$16\frac{5}{8}$	$5\frac{3}{4}$	$1\frac{1}{4}$	$\frac{5}{16}$	$6\frac{3}{8}$	3	$8\frac{3}{4}$	$5\frac{1}{4}$	15
18″	$7\frac{1}{8}$	$2\frac{11}{16}$	$18\frac{5}{8}$	$6\frac{1}{2}$	$1\frac{7}{16}$	$\frac{3}{8}$	$7\frac{1}{4}$	$3\frac{3}{8}$	$9\frac{1}{2}$	$5\frac{1}{2}$	$15\frac{3}{4}$
20″	8	3	21	$7\frac{1}{4}$	$1\frac{1}{2}$	$\frac{3}{8}$	8	$3\frac{1}{2}$	10	$6\frac{5}{8}$	$17\frac{1}{2}$
22″*											
24″*											
26″*											

* Thimble and link sets for these very large ropes are made to Individual Order and Specification

For cordage of up to nine inches in circumference . . . (and in this instance the term 'cordage' is used to cover all ropes whether of steel, fibre or man-made) . . . it is customary to line an eye-splice with a thimble only.   In cordage of over 9 ins. circumference, thimble and link sets are used, this is because the cross-sectional diameter of the larger sizes is such as to require joining shackles of such dimensions as to be practically unhandleable under ordinary conditions of sea towage.   Thimbles for use with tow ropes are made of mild steel which may be either left black or galvanised, cast hiduminium or cast bronze in special alloys.   The Links are forged in cable iron or mild steel to the appropriate British Institute Standards.

Joining Shackles.

The connection between the various parts of a towing set, as well as the connection to a towed vessel's cable, or to the bridle of any other floating object in tow, should always be made with a cable type shackle fitted with an oval sectioned bolt secured by the 'pin and pellet' method.   Shackles of the screw type should never be used because of a tendency, in this type, towards seized threads following upon the slightest distortion through stress.

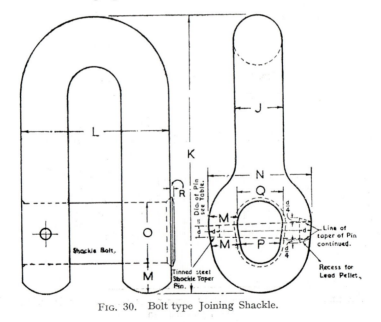

Fig. 30.   Bolt type Joining Shackle.

B.S. Ref. No.	J Diameter	K Length overall min.	K max.	L min.	M	N Width of shackle eye	O	P	Q	R	d Taper 1 in 16 on dia.
	in.	in.	in.	in.	in.	in.	in.	in.	in.	in.	in.
BS 189	$4\frac{7}{8}$	$26\frac{5}{8}$	$27\frac{3}{16}$	15	3	$10\frac{1}{2}$	6	$3\frac{3}{4}$	$4\frac{1}{2}$	$\frac{3}{8}$	$\frac{3}{4}$
BS 190	$4\frac{13}{16}$	$26\frac{3}{16}$	$26\frac{11}{16}$	$14\frac{3}{4}$	$2\frac{15}{16}$	$10\frac{5}{16}$	$5\frac{7}{8}$	$3\frac{11}{16}$	$4\frac{7}{16}$	$\frac{3}{8}$	$\frac{3}{4}$
BS 191	$4\frac{11}{16}$	$25\frac{3}{4}$	$26\frac{1}{4}$	$14\frac{1}{2}$	$2\frac{7}{8}$	$10\frac{1}{8}$	$5\frac{13}{16}$	$3\frac{5}{8}$	$4\frac{3}{8}$	$\frac{3}{8}$	$\frac{3}{4}$
BS 192	$4\frac{5}{8}$	$25\frac{5}{16}$	$25\frac{13}{16}$	$14\frac{1}{4}$	$2\frac{7}{8}$	10	$5\frac{11}{16}$	$3\frac{9}{16}$	$4\frac{1}{4}$	$\frac{3}{8}$	$\frac{3}{4}$
BS 193	$4\frac{9}{16}$	$24\frac{7}{8}$	$25\frac{3}{8}$	14	$2\frac{13}{16}$	$9\frac{13}{16}$	$5\frac{5}{8}$	$3\frac{1}{2}$	$4\frac{3}{16}$	$\frac{3}{8}$	$\frac{3}{4}$
BS 194	$4\frac{1}{2}$	$24\frac{7}{16}$	$24\frac{7}{8}$	$13\frac{3}{4}$	$2\frac{3}{4}$	$9\frac{5}{8}$	$5\frac{1}{2}$	$3\frac{7}{16}$	$4\frac{1}{8}$	$\frac{3}{8}$	$\frac{3}{4}$
BS 195	$4\frac{3}{8}$	24	$24\frac{7}{16}$	$13\frac{1}{2}$	$2\frac{11}{16}$	$9\frac{7}{16}$	$5\frac{3}{8}$	$3\frac{3}{8}$	$4\frac{1}{16}$	$\frac{5}{16}$	$\frac{3}{4}$
BS 196	$4\frac{5}{16}$	$23\frac{9}{16}$	24	$13\frac{1}{4}$	$2\frac{5}{8}$	$9\frac{1}{4}$	$5\frac{5}{16}$	$3\frac{5}{16}$	4	$\frac{5}{16}$	$\frac{3}{4}$
BS 197	$4\frac{1}{4}$	$23\frac{1}{8}$	$23\frac{9}{16}$	13	$2\frac{5}{8}$	$9\frac{1}{4}$	$5\frac{3}{16}$	$3\frac{1}{4}$	$3\frac{7}{8}$	$\frac{5}{16}$	$\frac{3}{4}$
BS 198	$4\frac{1}{8}$	$22\frac{11}{16}$	$23\frac{1}{16}$	$12\frac{3}{4}$	$2\frac{9}{16}$	$8\frac{15}{16}$	$5\frac{1}{8}$	$3\frac{3}{16}$	$3\frac{13}{16}$	$\frac{5}{16}$	$\frac{3}{4}$
BS 199	$4\frac{1}{16}$	$22\frac{3}{16}$	$22\frac{5}{8}$	$12\frac{1}{2}$	$2\frac{1}{2}$	$8\frac{3}{4}$	5	$3\frac{1}{8}$	$3\frac{3}{4}$	$\frac{5}{16}$	$\frac{5}{8}$
BS 200	4	$21\frac{3}{4}$	$22\frac{3}{16}$	$12\frac{1}{4}$	$2\frac{7}{16}$	$8\frac{9}{16}$	$4\frac{15}{16}$	$3\frac{1}{16}$	$3\frac{11}{16}$	$\frac{5}{16}$	$\frac{5}{8}$
BS 201	$3\frac{7}{8}$	$21\frac{5}{16}$	$21\frac{3}{4}$	12	$2\frac{3}{8}$	$8\frac{3}{8}$	$4\frac{13}{16}$	3	$3\frac{5}{8}$	$\frac{5}{16}$	$\frac{5}{8}$
BS 202	$3\frac{13}{16}$	$20\frac{7}{8}$	$21\frac{1}{4}$	$11\frac{3}{4}$	$2\frac{3}{8}$	$8\frac{1}{4}$	$4\frac{11}{16}$	$2\frac{15}{16}$	$3\frac{1}{2}$	$\frac{5}{16}$	$\frac{5}{8}$
BS 203	$3\frac{3}{4}$	$20\frac{7}{16}$	$20\frac{13}{16}$	$11\frac{1}{2}$	$2\frac{5}{16}$	$8\frac{1}{16}$	$4\frac{5}{8}$	$2\frac{7}{8}$	$3\frac{7}{16}$	$\frac{5}{16}$	$\frac{5}{8}$
BS 204	$3\frac{11}{16}$	20	$20\frac{3}{8}$	$11\frac{1}{4}$	$2\frac{1}{4}$	$7\frac{7}{8}$	$4\frac{1}{2}$	$2\frac{13}{16}$	$3\frac{3}{8}$	$\frac{5}{16}$	$\frac{5}{8}$
BS 205	$3\frac{9}{16}$	$19\frac{9}{16}$	$19\frac{15}{16}$	11	$2\frac{3}{16}$	$7\frac{11}{16}$	$4\frac{3}{8}$	$2\frac{3}{4}$	$3\frac{5}{16}$	$\frac{5}{16}$	$\frac{5}{8}$
BS 206	$3\frac{1}{2}$	$19\frac{1}{8}$	$19\frac{1}{2}$	$10\frac{3}{4}$	$2\frac{1}{8}$	$7\frac{1}{2}$	$4\frac{5}{16}$	$2\frac{11}{16}$	$3\frac{1}{4}$	$\frac{5}{16}$	$\frac{5}{8}$
BS 207	$3\frac{7}{16}$	$18\frac{11}{16}$	19	$10\frac{1}{2}$	$2\frac{1}{16}$	$7\frac{3}{8}$	$4\frac{3}{16}$	$2\frac{5}{8}$	$3\frac{3}{8}$	$\frac{5}{16}$	$\frac{5}{8}$
BS 208	$3\frac{5}{16}$	$18\frac{1}{4}$	$18\frac{9}{16}$	$10\frac{1}{4}$	$2\frac{1}{16}$	$7\frac{3}{16}$	$4\frac{1}{16}$	$2\frac{9}{16}$	$3\frac{1}{16}$	$\frac{5}{16}$	$\frac{5}{8}$
BS 209	$3\frac{1}{4}$	$17\frac{3}{4}$	$18\frac{1}{8}$	10	2	7	4	$2\frac{1}{2}$	3	$\frac{1}{4}$	$\frac{5}{8}$
BS 210	$3\frac{3}{16}$	$17\frac{5}{16}$	$17\frac{5}{8}$	$9\frac{3}{4}$	$1\frac{15}{16}$	$6\frac{13}{16}$	$3\frac{7}{8}$	$2\frac{7}{16}$	$2\frac{15}{16}$	$\frac{1}{4}$	$\frac{1}{2}$
BS 211	$3\frac{1}{16}$	$16\frac{7}{8}$	$17\frac{3}{16}$	$9\frac{1}{2}$	$1\frac{7}{8}$	$6\frac{5}{8}$	$3\frac{13}{16}$	$2\frac{3}{8}$	$2\frac{7}{8}$	$\frac{1}{4}$	$\frac{1}{2}$
BS 212	3	$16\frac{7}{16}$	$16\frac{3}{4}$	$9\frac{1}{2}$	$1\frac{7}{8}$	$6\frac{1}{2}$	$3\frac{11}{16}$	$2\frac{5}{16}$	$2\frac{3}{4}$	$\frac{1}{4}$	$\frac{1}{2}$
BS 213	$2\frac{15}{16}$	16	$16\frac{5}{16}$	9	$1\frac{13}{16}$	$6\frac{5}{16}$	$3\frac{5}{8}$	$2\frac{1}{4}$	$2\frac{11}{16}$	$\frac{1}{4}$	$\frac{1}{2}$
BS 214	$2\frac{13}{16}$	$15\frac{9}{16}$	$15\frac{13}{16}$	$8\frac{3}{4}$	$1\frac{3}{4}$	$6\frac{1}{8}$	$3\frac{1}{2}$	$2\frac{3}{16}$	$2\frac{5}{8}$	$\frac{1}{4}$	$\frac{1}{2}$
BS 215	$2\frac{3}{4}$	$15\frac{1}{8}$	$15\frac{3}{8}$	$8\frac{1}{2}$	$1\frac{11}{16}$	$5\frac{15}{16}$	$3\frac{3}{8}$	$2\frac{1}{8}$	$2\frac{9}{16}$	$\frac{1}{4}$	$\frac{1}{2}$
BS 216	$2\frac{11}{16}$	$14\frac{11}{16}$	$14\frac{15}{16}$	$8\frac{1}{4}$	$1\frac{5}{8}$	$5\frac{3}{4}$	$3\frac{5}{16}$	$2\frac{1}{16}$	$2\frac{1}{2}$	$\frac{1}{4}$	$\frac{1}{2}$
BS 217	$2\frac{5}{8}$	$14\frac{1}{4}$	$14\frac{1}{2}$	8	$1\frac{5}{8}$	$5\frac{5}{8}$	$3\frac{3}{16}$	2	$2\frac{3}{8}$	$\frac{1}{4}$	$\frac{1}{2}$
BS 218	$2\frac{1}{2}$	$13\frac{13}{16}$	14	$7\frac{3}{4}$	$1\frac{9}{16}$	$5\frac{7}{16}$	$3\frac{1}{8}$	$1\frac{15}{16}$	$2\frac{5}{16}$	$\frac{1}{4}$	$\frac{1}{2}$
BS 219	$2\frac{7}{16}$	$13\frac{5}{16}$	$13\frac{9}{16}$	$7\frac{1}{2}$	$1\frac{1}{2}$	$5\frac{1}{4}$	3	$1\frac{7}{8}$	$2\frac{1}{4}$	$\frac{1}{4}$	$\frac{1}{2}$
BS 220	$2\frac{3}{8}$	$12\frac{7}{8}$	$13\frac{3}{8}$	$7\frac{1}{4}$	$1\frac{7}{16}$	$5\frac{1}{16}$	$2\frac{7}{8}$	$1\frac{13}{16}$	$2\frac{3}{16}$	$\frac{3}{16}$	$\frac{3}{8}$
BS 221	$2\frac{1}{4}$	$12\frac{7}{16}$	$12\frac{11}{16}$	7	$1\frac{3}{8}$	$4\frac{7}{8}$	$2\frac{13}{16}$	$1\frac{3}{4}$	$2\frac{1}{8}$	$\frac{3}{16}$	$\frac{3}{8}$
BS 222	$2\frac{3}{16}$	12	$12\frac{3}{8}$	$6\frac{3}{4}$	$1\frac{3}{8}$	$4\frac{3}{4}$	$2\frac{11}{16}$	$1\frac{11}{16}$	2	$\frac{3}{16}$	$\frac{3}{8}$
BS 223	$2\frac{1}{8}$	$11\frac{9}{16}$	$11\frac{3}{4}$	$6\frac{1}{2}$	$1\frac{5}{16}$	$4\frac{9}{16}$	$2\frac{5}{8}$	$1\frac{5}{8}$	$1\frac{15}{16}$	$\frac{3}{16}$	$\frac{3}{8}$
BS 224	2	$11\frac{1}{8}$	$11\frac{5}{16}$	$6\frac{1}{4}$	$1\frac{1}{4}$	$4\frac{3}{8}$	$2\frac{1}{2}$	$1\frac{9}{16}$	$1\frac{7}{8}$	$\frac{3}{16}$	$\frac{3}{8}$
BS 225	$1\frac{15}{16}$	$10\frac{11}{16}$	$10\frac{7}{8}$	6	$1\frac{3}{16}$	$4\frac{3}{16}$	$2\frac{3}{8}$	$1\frac{1}{2}$	$1\frac{13}{16}$	$\frac{3}{16}$	$\frac{3}{8}$
BS 226	$1\frac{7}{8}$	$10\frac{1}{4}$	$10\frac{3}{8}$	$5\frac{3}{4}$	$1\frac{1}{8}$	4	$2\frac{5}{16}$	$1\frac{7}{16}$	$1\frac{3}{4}$	$\frac{3}{16}$	$\frac{3}{8}$
BS 227	$1\frac{3}{4}$	$9\frac{13}{16}$	$9\frac{15}{16}$	$5\frac{1}{2}$	$1\frac{1}{8}$	$3\frac{7}{8}$	$2\frac{3}{16}$	$1\frac{3}{8}$	$1\frac{5}{8}$	$\frac{3}{16}$	
BS 228	$1\frac{11}{16}$	$9\frac{3}{8}$	$9\frac{1}{2}$	$5\frac{1}{4}$	$1\frac{1}{16}$	$3\frac{11}{16}$	$2\frac{1}{8}$	$1\frac{5}{16}$	$1\frac{9}{16}$	$\frac{3}{16}$	$\frac{1}{4}$

CHAPTER II.

Splicing Methods Used on Board of Ocean-Going Tugs.

Although new tow ropes and towing springs are invariably supplied with the necessary eye-splices already formed, it is frequently necessary to do this on board of the tugs. This is occasionally called for at sea in emergencies following the parting of gear, but is called for much more often when towing sets are required for special towage operations, or in the case of making up selections of smaller towing sets required by the various classes and types of tug according to their conditions of service and employment.

1. Eye-Splices in Manila or Sisal Towing Springs.

The splices used in these ropes are entirely straightforward, exactly the same methods being used as in any other fibre ropes, the only complications arising being those resulting from the rather over-sized dimensions of such ropes.

This difficulty is however successfully overcome by the use of splicing tools of proportionate sizes. The fids used in such work being, for instance, some four feet in length, tapering from about ten inches in diameter at the butt down to a blunt point. Because these fids are used in conjunction with heavy mallets, their butts are made quite flat and are fitted with steel ferrules. It is also common practice among tug seamen to make recourse to a messenger from the capstan when rousing through tucks in a stiff rope.

Four full tucks are commonly taken in eye-splices for towing work, the strands being middled and dogged after the last tuck. I will not dwell here upon procedural detail seeing that this is more than adequately dealt with in every standard work on seamanship. One advice only is tendered towards the amelioration of physical difficulties.

Some springs are extremely hard laid, particularly sisal; this makes it more than usually difficult to enter the fids when making tucks. To facilitate the business, proceed as follows. When the spring has been set up with everything ready for splicing, put a strop of $3\frac{1}{2}$ ins.–4 ins. fibre rope around the spring some eight to ten feet away from the thimble. Pass the strop

86

through its own part about the spring and take a few turns around the spring against the lay leaving enough bight to take a hand-spike or capstan bar. Using this as a lever, take a half turn out of the spring and then lash it *in situ*. This amount of opening of the rope allows of a much easier penetration by the fid and sensibly eases the passage of the strands whilst tucking. The lever may be eased back upon the completion of each round of tucks in order that the splicer may fully settle the strands.

2. Eye-Splices in Flexible Steel Wire Rope.

Steel wire tow ropes, in common with all ropes laid up in the rotational principle, tend to unlay when subjected to stress. Eye-splices which incorporate tucks taken *with* the lay are therefore never used in forming eyes in towing wires because of the risk of them drawing under the influence of this unlaying process. Splices made in the 'Liverpool,' 'French' or 'Spiral' fashion are therefore never used. All eyes in towing wire ropes are formed with splices which use tucks taken *against* the lay, and it is best tug practice to utilise a style which provides a locking principle.

Before entering into the detail of a description of a typical towing splice, the need for a properly fitted rigger's vice must be indicated. Towing wires, although flexible, are always on the large side and are always extremely tough. They cannot be properly set up for splicing in an ordinary vice, and the assistance of the crown screw in the rigger's vice is entirely essential to a satisfactory job. Some splicers favour chisel ended spikes, some prefer the true point, and many like a combination of both, but regardless of the type, they must be clean, straight and sharp with well finished butts.

The splice to be described is a perfectly straightforward against the lay splice. There are at least three different variations of this, with different procedures resulting in detail differences in locking effect, but all come out with each strand emerging from a separate interstice after which each is tucked over one and under one against the lay. (Figure No. 31).

Commence operations by putting on a good seizing at the appropriate distance from the end of the wire; then open up the wire carefully and whip the ends of the strands, if the necessary heat is available, weld them up and grind smooth, it saves time in the end. Seize in the thimble at the proper place and set it all up tightly in the riggers' vice. Arrange the wire so that the crown is towards you with the unlayed end lying on the *left* side of the standing part. Remove the heart to the seizing, middle the strands and lay

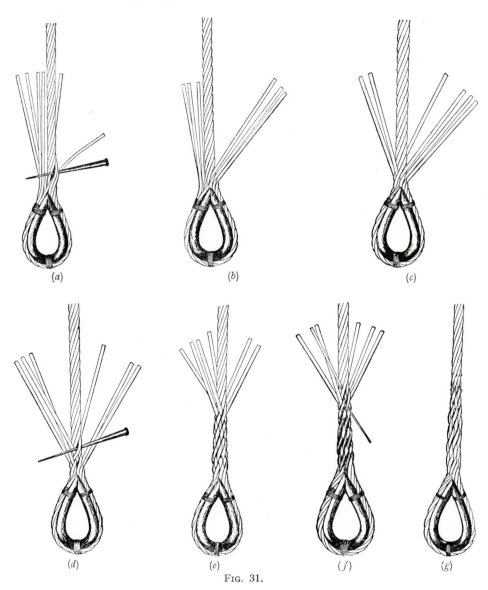

(a)   (b)   (c)

(d)   (e)   (f)   (g)

Fig. 31.

them down beside the standing part. For convenience in description, the strand nearest to the standing part and on the top is No. 1, the others following successively Nos. 2, 3, 4, 5 and 6.

Insert the spike from right to left under two strands slightly above the throat of the thimble. Tuck strand No. 1 under the spike and under both strands and from left to right. Pull this strand well home using a smooth round headed hammer if necessary. Withdraw the spike and re-insert it one strand to the left of the interstice from which strand No. 1 emerges, that is the leftmost strand of the two that were initially lifted. Tuck strand No. 2 under this and pull it well home. Withdraw the spike and insert it under the next strand again to the left, pick up strand No. 6 and tuck it underneath this, again from left, do not now withdraw the spike but force it further in, in order to enlarge the interstice because strand No. 3 is tucked from right to left under this strand. Here now take the greatest care to set both of these strands up taut before withdrawing the spike. Insert the spike under the next strand to the left and tuck No. 4 similarly. Repeat with No. 5.

This is the first tuck, with lock completed and the splice should be taken down out of the riggers' vice and settled properly by hauling the first tuck well up to the throat of the thimble using the round headed hammer as necessary. It is quite impossible to spend too much time on settling the first tuck.

Return the splice to the riggers' vice. Begin the second tuck by taking up strand No. 6. Insert the spike under the second strand to its left in the standing part. Take No. 6 over the intervening strand and in under the spike from right to left and against the lay of the wire. Repeat this procedure with the other five strands in order viz. Nos. 3, 4, 5, 1 and 2. This completes the second tuck. Third and fourth tucks being taken in exactly the same manner 'over one and under one,' working from right to left and against the lay, the splice being removed from the vice in between each to thoroughly set the the tuck, using the hammer as necessary.

The fifth tuck is the first half tuck and is ordered by reducing the wires in each strand by a half before taking them over one and under one.

The sixth tuck, and the second half tuck is achieved by tucking strands Nos. 1, 3 and 5 only, and these over one but under *two*. This provides a fully effective taper besides providing the best possible lock to the strand ends.

It is the practice in some ocean-going tugs always to splice soft eyes into their steel wire tow-ropes, and then to secure the thimble therein by means of a close-up throat racking seizing made with $\frac{1}{2}$ inch flexible steel wire.

There is positively no objection to this practice so long as the seizing is properly made and finished; indeed there is a certain advantage thereby deriving when a thimble becomes worn before the wire has suffered deterioration and so can be replaced without the necessity of a new splice.

In the matter of services for these splices, there has always been two sharply differing opinions. One holds to the necessity of a wire service over every splice because of the support thus provided and because the service provides a smooth and snag-free hand-hold. The other contends that these splices should be left naked so as to facilitate inspection. This school contends too that because of inevitable scarring and loss of surface finish, all splices are vulnerable to corrosion, a circumstance which is assisted by such water-retaining material as is provided by the servings. Of the two views, possibly the former has the greater force.

One final word. Correct tapering sensibly reduces the loss in strength in steel wire ropes which is associated with all splicing. *This loss can amount to 5 per cent, or even more, when skill in splicing is wanting.* An effective taper, formed as in the foregoing description, not only keeps strength loss to a minimum but also sensibly assists in the subsequent handling of the rope by virtue of the flexibility of the rope, right up to the thimble.

3. Eye-Splices in Ropes Made from Man-made Fibres.

Towing springs made from Nylon, Terylene, Polythene and Polypropylene are all of the plain right-handed hawser-laid style and present no difficulty whatsoever. When length is no problem, springs are made to the appropriate length from suitable sizes of rope, and a thimble is worked into an eye at both ends in the conventional way. When, however, it is necessary to reduce length, it has become the practice to make up a two part spring by forming a strop to the desired length and then to seize a suitable thimble into each end by means of a proper series of seizings.

In both cases traditional against the lay procedures are used, the only difficulties obtaining being those which derive from the smooth and silky nature of the filaments.

Care must be taken in unlaying rope of this type so as not to lose the shape of the lay and because of this delicacy and silkiness more tucks are required than with natural fibre. Six are commonly taken and the practice of middling and dogging the strand-ends is utilised as with fibre ropes.

4. Bull-Dog-Grips.

A section upon splicing methods cannot be terminated without making reference, in connection with steel wire ropes, to the ubiquitous bull-dog-grip. This fitting is very much disparaged by professional seamen of all nationalities, probably because of the appearance of eyes and loops formed with their assistance but, if one can accept that these grips are only for use in an emergency and only when there is no time for a splice, then their efficiency and utility can be admired if their appearance can only be deplored.

Bull-dog-grips to B.S.I. standards comprise an 'U' bolt and nuts and washers and a bridge piece. The 'U' bolt must be bent to shape when heated to a uniform red heat, the twin legs, of equal length, thereafter being threaded to the appropriate gauge Whitworth and provided with hexagonal nuts and spring washers.

The bridge piece is either a steel forging or a steel casting and must be grooved on its inside bearing surface in precise replica of the steel wire rope with which it will be associated. The bolt holes must be cored or bored to provide a free and parallel fit for the 'U' bolt legs.

Not less than three grips should be used to form an eye or loop in steel wire rope and in ropes of from 3 ins. to 4 ins. circumference, four grips must be used. Ropes of more than 4 ins. circumference require five grips.

These grips must however be fitted correctly. The bridge part of the fitting must locate upon the working part of the wire with the 'U' bolt bearing upon the lazy or tail end of the rope. The grips should be spaced at from 9 ins. to 12 ins. apart according to the size of the wire with the first one as close as can be contrived to the throat of the thimble.

CHAPTER III.

Combination Towing Media.

It is commonly the case that a towing medium will consist of a combination of steel wire rope together with a fibre or synthetic filament spring, and in this there is a condition of more than passing importance.

Nearly all cordage for towing is made up right-handed, that is, when a section of the cordage is held out straight before an observer the strands incline to the right. *But, not all of cordage which is suitable to towage use is so constructed, and there is a deal of cordage in use in other countries which is of left-handed construction.*

Now, there is no especial advantage accruing in having rope laid up right-handed as opposed to left-handed or vice versa, but it is important that all cordage in any one tug should be one or the other, *but never both*, this is for a very good reason.

Any rope suffering stress tends to unlay a little. Any rope being towed through the water also tends to revolve under water effect upon the lay in a direction opposite to the lay. It follows then that, because the greater part of ocean towing gear is submerged and is under stress it must suffer a tendency towards unlaying in the sum of these two effects. It is important here to say that a little unlaying of rope is not harmful and is allowed for by the manufacturer when the initial torque is introduced, but excessive unlaying is obviously disastrous.

When a towing combination of wire and fibre, or synthetic ropes is all of right-hand lay then there is no interruption of the forces resolving throughout the combined length, and the effect is, for all practical purposes, undiscernable, the total length of gear being such as to absorb the unlaying processes without ill-effect. When a combination is of opposite hands is in use however, this produces opposite forces which meet in competition at the point of connection. When it is a combination of manila and steel wire and the two parts are of the same length, one usually finds that the manila triumphs over the steel wire rope largely, it is believed, because of the leverage offered by the width of the thimbled eye. This causes the steel wire rope to unlay about the splice, with obvious ill-effect. When the fibre is shorter than the steel wire component, or is in poor shape, the wire will open up the manila, also

92

with deleterious result. This ill-effect is inversely proportional the the separate lengths of media in use, viz. the shorter the component the greater the effect. This becomes most noticeable in docking springs. It has been known, for instance, for a 10 inch cable laid manila of left-hand lay to unlay a 3 inch flexible steel wire pennant of right-hand lay over a length of two fathoms from the connection. Similarly, a 3 inch right hand laid flexible steel wire pennant has been known to unlay a 14 inch grass spring to such an extent as to make it useless. The same effects obtain with springs of synthetic materials.

The Towing Equipment Carried On Board of the Various Classes of Sea and Ocean-Going Tugs.

The proper and most suitable sizes and types of towing gear to be supplied to the various classes of tug has always seriously exercised the considerations of all interested parties ashore and afloat. There is, quite naturally, a very great deal of technical and professional controversy in this matter, a condition which has not been eased since the advent of man-made fibres on to the towage scene, largely because of the comparisons which are inevitably made.

Whilst it is clearly quite uneconomical to send tugs away to sea equipped with towing gear substantially in excess of practical requirements, it is obviously even less economical to equip a tug in such a fashion that she is unable to exert her fullest potential on towage operations.

There have, from time to time, been efforts made to rationalise the issue by relating tow-rope strength to the bollard pull achieved by tugs. The immediate difficulty arising having been the selection of an appropriate standard of strength. There has always been too the thought that towing gear, under conditions of adverse weather, sometimes has to accept stress in excess of the tug's bollard pull. In the endeavours towards rationalisation, the advice of rope manufacturers was sought. In the light of laboratory experiment and the vast resources in terms of record available to them, the manufacturer submitted that, whatever relative factor in terms of power available was used, the factor for the fibre and steel wire tow ropes, in terms of a proportion of ultimate strength available for constant working should not exceed 25 per cent. In the case of man-made fibres, they considered that here 50 per cent of the ultimate strength should be the maximum constant stress which might safely be supported.

The manufacturer was, in the opinion of most seamen, courageous indeed

in coming out with so firm a statement and it occasioned no comment whatsoever when the manufacturer offered that the stress stipulated was that of pure
extension in terms of suspended weights and could not take into account
such other stress as may be attendant such as vibration, crushing stress and
torsional stress.

No one who has placed his hand on a tow-rope whilst towing could ever
deny the presence of vibrational effect. It is also entirely evident that there
is a deal of crushing stress in the locality of a tug's taffrail under every condition of ocean towing. It is apparent that crushing stress must also obtain
upon the turns taken in wire on towing bollards, in the coils upon a towing
winch barrel, over leads and certainly about the eye-splices. All of this sort
of stress being very much accentuated under conditions of adverse weather.

Whilst, in the case of fibre and steel wire ropes, the 25 per cent rule seems
fair in the smaller classes of tug, it clearly becomes somewhat impracticable
once one is in consideration of the bollard pulls deriving from Brake Horse
Powers of 5,000 and upwards. In all, practical experience afloat would
appear to indicate that the correct size of gear to be carried in any class of
tug is . . . *'The largest rope, of the finest quality, which can be safely manipulated
by the personnel of the tug, taking into consideration the space available and the
adequacy, or otherwise, of the tug's fittings in terms of capstans, winches,
bollards, leads and so on'* . . . In the case of the more modern man-made
fibres, however, one may approximate to the 50 per cent rule in view of the
markedly superior performance of such ropes. This resolution of the problem
will not, of course, receive the approval of the more technically minded of
the interested parties, but it is certain to occasion the approval of tug
personnel which is possibly more important.

In the practical application of this consideration it is perhaps convenient
to divide sea and ocean-going tugs into four broad classes:—

1.—The Ocean Class.

This class should include tugs of 5,000 B.H.P. and upwards. Tugs
deemed capable of operating in the open ocean in all weathers and of handling,
single handed the largest passenger ships, tankers and bulk carriers.

2.—The Deep-Sea Class.

This class should include tugs of from 3,000 to 4,000 B.H.P. Tugs
deemed capable of operating in the open ocean in all weathers and of handling
single handed vessels of up to 45,000 tons displacement.

3.—Sea-Going Tugs.

This class to include tugs of up to 2,500 B.H.P. Tugs capable of operating in the open seas in all weathers and of handling single handed vessels of up to 30,000 tons displacement.

4.—The Coastal Class.

Tugs of up to 1,500 B.H.P. Such tugs being considered capable of operating on towage and rescue duties in coastal waters under weather conditions of normal native severity.

Note:—Tugs of classes 1, 2 and 3 will invariably be fitted with a single or duplex towing winch. The coastal class will only occasionally be provided with this auxiliary.

1. Equipment for an Ocean Class Tug.

(*a*) 2×450 fathom E.S.F.S.W. 6×61 towing wires mounted in twin drums on the towing winch of $6\frac{1}{2}$ inch—$7\frac{1}{2}$ inch circumference. Each wire to be fitted with a thimbled eye in one end with the other ferruled or tapered . . . (Advantage may well derive from having one of these wires somewhat smaller than the other).

(*b*) Two similar wires carried as spares.

(*c*) 2×120 fathom cable laid manila towing springs of 24 inch—26 inch circumference, fitted thimble and link at each end* or 2×90 14 inch nylon springs similarly fitted or 2×60 double 12 inch stropped).

(*d*) A selection of $6\frac{1}{2}$ inch—$7\frac{1}{2}$ inch pennants of 60 to 90 fathoms length to use with (*c*).

2. Equipment for a Deep-Sea Class Tug.

(*a*) 2×450 fathom E.S.F.S.W. 6×61 towing wires mounted in twin drums on a towing winch of $5\frac{1}{2}$ inch—6 inch circumference. Each wire to be fitted with a thimbled eye in one end with the other ferruled or tapered . . . (Advantage might well derive from having one of these wires smaller than the other).

(*b*) One spare winch wire (or two if different size wires are carried).

(c) 2×120 fathom cable laid manila towing springs of 20 inch—22 inch circumference, fitted thimble and link at each end* or 2×90 12 inch nylon springs similarly fitted or 2×60 double 9 inch stropped.

(d) A selection of $5\frac{1}{2}$ inch—6 inch pennants of 60 to 90 fathoms to use with (c).

(e) $2 \times$ Close quarters handling springs made up of $4\frac{1}{2}$ inch—5 inch E.S.F.S.W. and 16 inch manila or 10 inch Terylene. (These springs are especially made for each tug. The Fibre or Terylene Spring being made to a length very slightly greater than the distance from the point of tow to the taffrail with the pennant about 20 fathoms long. Thimbles are used to line the splices in the spring section but none are used in the eye of the pennant. The pennant being spliced directly into the fibre thimble. This in the interests of lightness and handiness).

Being made to measure these springs are passed easily and the reduced dimensions of this equipment is permissible in view of the fact that it will be used in smooth water.

3. Equipment for a Sea-Going Tug.

(a) 2×450 fathom E.S.F.S.W. 6×61 towing wires mounted in twin drums on a towing winch of $4\frac{1}{2}$ inch—$5\frac{1}{4}$ inch circumference. Each wire to be fitted with a thimbled eye in one end with the other ferruled or tapered.

(b) One spare winch wire.

(c) 2×120 fathom cable laid manila towing springs of 18 inch circumference, fitted thimble and link at each end or 2×90* 10 inch nylon springs similarly fitted or 2×60 double 7 inch stropped.

(d) A selection of $4\frac{1}{2}$ inch—$5\frac{1}{4}$ inch pennants of from 45 to 90 fathoms to use with (c).

(e) $2 \times$ Close quarters handling springs of 4 inch E.S.F.S.W. and 14 inch manila or 8 inch Terylene fitted as described for the Deep-Sea Class.

4. Equipment for the Coastal Class.

(a) 2×120 fathom cable laid manila towing springs of 14 inch—16 inch circumference, fitted thimble and link at each end or 2×90 fathom 8 inch or 9 inch nylon springs similarly fitted.

(b) 2×75 fathom lengths similarly fitted.

(c) 2×50 fathom lengths similarly fitted.

(d) A selection of E.S.F.S.W. pennants of $3\frac{1}{2}$ inch—4 inch circumference in varying lengths up to 90 fathoms.

(e) Close quarters working gear in the form of harbour springs would naturally form part of this class's basic equipment as they would ordinarily be employed on river or estuarial towage. The style and quantity of this gear would depend upon the local requirement.

Note 1.—

All classes of tugs would carry a selection of spare thimbles, thimbles and links and shackles.

Note 2.—

The shorter hawsers and more diverse selection of gear generally in the Coastal class is due to the fact that these craft carry out their sea duty in much more circumscribed conditions than is the case with their bigger sisters. No more accurate account can be offered than that given because all will largely depend upon the sphere of operations allotted to particular tugs. A tug based at Hull, for instance, would, in all probability, find the major part of its work resulting from strandings on the sandbanks off the East Anglian Coasts or in short tows resulting from collision casualties in the restricted channels adjacent thereto. Conditions for a tug based in the Clyde being similarly conditioned by local geography.

Note 3.—

The prudent Commanding Officer of one of the larger classes would almost certainly provide himself with a modest quantity of studded link cable for use in the construction of bridles or stranding operations.

Note 4.—

The items marked * would be a special order in the United Kingdom and might well be unobtainable in many foreign countries.

Note 5.—

All of the foregoing is offered in terms of 'average usage' it being entirely understood that many tug operators have the strongest views upon scales and details of equipment. It will clearly require amendment in the case of the larger classes where either single barrel winches are used or where no winch is fitted at all. In any case the detail offered should be regarded as a minimum.

CHAPTER IV

The Stowage, Care and Maintenance of Ocean Towing Equipment

Besides the ordinary day to day cordage which must be carried by any and every vessel sailing the high seas and which, of itself, represents a capital outlay of no mean proportions, the ocean-going tug requires an outfit of specialised professional cordage which may well be valued in terms of thousands of pounds. Not only does she require the basic items of equipment in terms of towing springs and wires, but also the manifold lesser, but nevertheless vital, items of auxiliary equipment such as messengers, heaving lines, rocket lines, beaching and floating lines, bridles, gog-ropes and so forth. So that these may be available as required it becomes necessary to carry quite consequential quantities of the various sizes and types of ropes so that it is clearly obligatory upon tug officers to achieve expertise in the care, maintenance, stowage and proper selection and use of the variety of gear to hand.

Every large tug capable of operating independently of a base will certainly be provided with a hold, or large stores space, where spare gear may be safely stowed and it would be the concern of any responsible officer to see that this space was properly arranged and organised to the purpose. His first concern would naturally be for the essentials, the need for shelving, gratings, etc., being secondary to the absolute necessity for the space being quite dry and well ventilated.

In most large tugs the storage space takes the form of a hold below the towing deck, and it is customary to provide a heavy grating, raised a few inches above the ceiling, to occupy the whole of the square below the hatch. Other gratings usually obtain in shelf form, two or three deep in the wings of the hold whilst a further two or three are often disposed athwartships across the after part of the hold. There will, of course, be other stowage arrangements in the form of reels for the smaller stuffs, and bins for falls and messengers and the like, depending upon the views and requirements of the operator and the ingenuity of the responsible officers.

It is, in most tugs, the practice to coil the spare fibre or nylon springs down on the after half of the central grating. When only one spring is carried, it is common practice to bring one end out clear and trice it up to

98

an eye bolt on the after coaming by a lashing into the thimble link to be readily accessible in an emergency. When more than one is carried, the first reserve is similarly arranged. Where a selection of short springs is carried for use with a towing winch, it is customary to bring an eye of each out clear and laid aft and marked with a tally. Pennants for use with these springs are always stowed clear and are similarly identified. For very obvious reasons all gear is kept separate and is lashed.

Spare towing winch wires are usually flaked down in large coils on the fore side of the grating with the tapered or ferruled end triced up to the fore-coaming. Though it is difficult to see why it is not possible to stow them on reels suitably worked into the tug's fabric in such fashion that they can be passed directly on to the winch barrel through a suitable aperture.

Close quarters gear and docking springs are usually found in the wing gratings where care is taken to keep the steel wire pennants clear of the fibre or synthetic spring sections. The selections of steel wire pennants are usually flaked down in easy coils on the after part shelves and it is best tug practice to mark these with wooden or canvas tallies detailing their lengths and circumferences. All of the coils of smaller stuffs whether of steel wire, fibre or synthetic are commonly divided into two categories, namely ready use and reserve, and are prominently marked accordingly. In certain well-found tugs, special reel fittings are worked into the shelving arrangements for the proper accommodation of the ready use gear.

It is increasingly the case that the larger classes of tug carry small quantities of studded link cable in sizes appropriate to the rest of the gear for the purpose of making up towing bridles for special operations or for use in salvage operations. The length carried varies considerably according to the operators' notions in the matter, but ordinarily consists of fifteen to thirty fathoms.

To ensure efficient stowage, all gear should be dry, clean and free of grease or oil. It is realised that the first requirement, in terms of actual operating conditions, will often be difficult of achievement, but it is a condition worth striving for if gear is to be maintained in top condition. Routine inspections of all gear stowed below decks is quite essential, particular attention being paid to evidence of rope-rot or mildew as evidenced by the associated unpleasant odours and dark brown staining.

Spare tow-rope accessories such as thimbles, thimble and link sets and shackles should be stowed together as convenience dictates. There is no objection to a coat of paint on any of these items, indeed galvanised primer type paint is ideal for the purpose, but this should not obtrude upon the fitted bearing areas in shackle bolts and pins. These areas do, of course, require protection but this is better provided by a thin smear of grease or blue mercurial ointment.

New fibre gear should always be coiled down as well as initial stiffness will permit, but at the first opportunity it should be streamed astern and thoroughly stretched, after which it should become sufficiently pliant to be properly put down. Avoid temptation to put tow-ropes down in flemish coils; such coils have an extremely neat and seamanlike appearance and they certainly conserve space but they are really only suitable for ropes which have, in use, both ends free to allow turns to run out. Tow-ropes in use are secured at both ends so that the turns put into them, when forming flemish coils, cannot run out and accumulate instead to form harmful kinks.

All rope suffers from contact with, or even close proximity to, all acids and alkalis, paint, varnish, mineral oils and greases and any of the petroleum derivatives in common use at sea. Harmful effect derives from dampness, excessive heat or cold and even exposure to strong sunlight. It is therefore prudent to keep all gear covered, whether on deck or below, with effective covering material of adequate strength to the purpose.

It is, however, entirely obvious that some towing gear must always be stowed so as to be immediately available, especially in the case of tugs not fitted with towing winches. Where a winch is fitted it is a great convenience to have one short pennant and spring stowed on deck, whilst in tugs not fitted it is essential to keep one complete towing set together with one pennant at first readiness on deck. In each case the gear should be stowed under a heavy canvas cover and either wholly within a proper container or upon a proper grating raised some inches above deck level. This is especially the case when the springs are constructed from one or other of the man-made fibres where there is a rather special vulnerability to the effects of strong sunlight and physical damage.

Under tropical conditions fibre towing springs tend to dry out and become brittle, particularly if stowed on deck. The natural oil latent in the fibre gradually dries out under such conditions, as does also the patent preservative dressing applied by the manufacturer. This is a circumstance which cannot be totally averted by even the most conscientious maintenance, but vigilance and the timely application of preservative compounds can certainly postpone the end-effects of such conditions. In the event that patent preservatives are not available, a fair practicable substitute is a periodic very light dressing with warmed vegetable oil.

In tropical circumstances too there is a tendency for the preservative and lubricative dressings on steel wire ropes to become progressively less viscuous and to run to the bottom of the coil. Spare gear must, therefore, be turned over from time to time, whilst the winch wire in use will obviously profit from a periodic 180° of rotation. Whilst discussing this aspect of maintenance, it is convenient to offer that all steel wire rope requires a periodic freshening of protective dressing. In the case of the winch wire this is best applied when paying out or heaving in, but other lesser lengths must be taken out of storage coils for this purpose. Here again the best dressing is the patented compound designed expressly for the job, with a second best comprising one or more thin coats of boiled linseed oil applied when hot.

After a consideration of tropical effect it would appear to be entirely logical to give thought to the opposite extreme. Experience of sub-polar working in two world wars has shown that wet fibre ropes can freeze under the appropriate conditions, and that when they are frozen they become so brittle as to be virtually useless. Clearly there are no palliative or curative measures to apply under these extreme conditions and the only profit deriving from this information lies in the fore-warning thus provided that such catastrophe can arise. This does not, however, occur in the case of ropes made of synthetic materials; although such ropes do take on a film of ice under the appropriate conditions this is immediately shed when the rope flexes under stress.

The Survey of Towing Gear.

Having made full reference to the care and maintenance of towing gear, all of which has as its ultimate aim the prolongation of life, and preservation of efficiency in the various types of towing media, it becomes necessary to give thought to a circumstance of considerable interest and importance to every tug officer; this is the consideration involved in reaching decisions as to the fitness, or otherwise, of suspect items of towing gear to remain in use. In this it is entirely proper to observe that it is one matter for a cargo ship's officer to condemn a derrick guy pennant or a cargo runner where, at best, the outlay involved is a matter of a few pounds only, but entirely another consideration when a tug officer is obliged to pass judgement upon 450 fathoms of E.S.F.S.W., for instance, or perhaps 120 fathoms of 20 inch manila, where replacement costs are measured in thousands of pounds. In the interests of sensible economy no responsible offiicer would wish to condemn unnecessarily such expensive equipment, but at the same time, the same officer dare not hazard the success of a towage operation by using suspect gear

H

The first requirement when carrying out a survey of any tow-rope must be that the rope in question must be fully exposed and entirely accessible throughout its length. This pre-supposes ample space, a condition which may not be obtainable in any but the very largest classes of tug. It is, therefore, almost invariably the rule that ropes for survey should be landed where they can be effectively man-handled to permit of the searching inspection which is indicated.

The principal causes of failure in all tow-ropes whether they may be of fibre construction, steel-wire or nylon or other man-made filament material, may be enumerated thus:—

> (a) Deterioration due to stress.
> (b) Physical damage.
> (c) Decay or corrosion.
> (d) Age.

1. Survey of Fibre Springs.

With the rope fully exposed, consider first the cause (a) by selecting a section of the rope which has suffered the least possible physical damage so that evidence as to stress failure shall not be obscured. First, carefully measure the rope's circumference and compare with the original dimension. In this initial test, however, it is important to remember that stretch in any rope under stress is quite inevitable and is in fact allowed for by the manufacturer who makes all fibre gear slightly oversize of standard. Initial stretch, therefore, is not harmful, but any subsequent stretching of a thoroughly settled rope must be taken as evidence of deterioration.

A second check on stretch is provided by measuring the angles of lay of the whole spring and of the three individual ropes forming the whole, using a large adjustable carpenter's bevel for the purpose. The proper angles of lay are 37° for the whole rope and 31° for the individuals. In both of these tests the surveying officer is looking for appreciable diminution from standard. In the matter of circumference for instance a reduction of 5 per cent would be the maximum acceptable whilst angles of lay should not suffer any greater diminution than 7 per cent. Any marked decrease in the lay of the whole rope in comparison with that of its three components should be taken as evidence of excess stretch.

The surveying officer, having completed these tests, should pass on to an examination of the interior of the rope, the necessary access being obtained by the utilisation of the strop and lever procedure described in the section on splicing. This operation of itself, will provide evidence of a rope's condition in relation to the degree of effort involved. With the interior of the rope exposed consider the condition of the individual ropes. Stress in ropes occasions compression besides elongation. The compression distorts the individual ropes from their original circular section to something approximating to the triangular. The yarns forming the strands do not all fairly bear the load thereafter causing some of them to sink within the strands or to break.

Next consider the cause (b) by thorough-footing the whole rope to find the worst damaged sections. This damage by either chafe or other abrasion represents lost strength over and above that found at (a). It naturally follows that most of this damage will have been inflicted upon the outside yarns of the strands, except that it, in the case of fracture or serious impact damage which must be separately assessed. It is therefore the outer and second layers of yarns which absorb most of the punishment but it must be clearly understood that a large proportion of all of the yarns in a strand *are* outers and seconds. Make a careful count of the damaged and broken yarns in one complete lay, add those deriving from the cause (a), and using the table appended hereafter, assess the proportion of whole yarns remaining. If 75 per cent of these are intact over the worst areas, then the rope may remain in service . . . if there are no other critical arisings under causes (b), (c) and (d) . . .

In the case of localised part fracture or severe damage the criterion must be the same with the proviso, however, that the damage may be cut out and the rope re-made with a short splice if the resultant loss of strength is acceptable.

Whilst the interior of the rope is exposed in examination of stress effects, the surveying officer should also make his appreciation of the incidence of decay. Decay is proportional to age assuming average standards of care and stowage, but the best of these cannot entirely arrest the drying out of natural oils and manufacturer's preservatives which obtains throughout the life of ropes. This progressive process allows ropes to absorb and retain more and more moisture as time goes by and, seeing that most of the moisture associated with ocean towage is sea water, ropes become impregnated with salt crystals, so that they become hygroscopic, a condition provocative of decay.

Colour and touch are the prime guide in this assessment. New fibre is yellow to brown yellow, depending upon its origin, but as it deteriorates, it first of all bleaches to a dirty white and then degenerates through the shades of grey to black. Test individual rope threads against new ones to ascertain relative flexibility and strength.

Finally, consider the history of the rope. In this it is offered that the initial cost of fibre-springs alone, not to mention the importance of its function, indicates the necessity for the maintenance of a history sheet for every spring in use. In this, besides age, due account should be paid to the sort of loading to which the rope has been subjected. It being apparent that a series of shock loadings over a short period, such as may be experienced on a reflotation after stranding operation, may well have a more destructive effect than the more even stress imposed by an ocean tow of months' duration, if this takes place in good weather, so that the rope is only being exercised well within its capacity.

Having carried out a survey along the general lines indicated, the officer should evaluate his findings remembering to keep it clear in his considerations that all evidence of deterioration is cumulative. There can be no rigid rules in this, but if a conscientious officer will follow this guide his opinion, properly expressed, can hardly be denied.

2. Steel Wire Towing Ropes.

Surveying procedures exactly as with fibre ropes except that the factor at (c) becomes corrosion instead of decay.

Stress deterioration indicates in steel wire ropes, as with any others, initially in terms of stretch, which the manufacturer again anticipates by making his ropes approximately 5 per cent oversized. Clearly, however, stretch in steel wire ropes is fractional in comparison to that in fibres so that one cannot readily assess it by the measurement of circumferences and angles of strand lay. It was observed earlier that the elongation effects produced by stress are accompanied by internal compression and this, in the case of steel wire rope provides the evidence of deterioration.

New steel wire ropes are provided with an extremely tough and flexible fibre heart of such diameter as to support the six strands of the rope with a minute gap between them so as to avert auto-abrasion. The first examination should ascertain whether this condition still obtains. After this the rope may be opened up to examine the condition of the heart utilising strop, lever and spikes, in such a manner as not to damage either the strands or the heart. It is not anticipated that any rope which has seen sufficient service as to justify surveying will contain a heart which is still of true round section and some degree of grooving is to be expected. It should not, however, be so distorted as to allow the wire strands to bear substantially, one upon the other. It should also still contain sufficient of the original impregnation of lubricant cum preservative to enable it to function effectively.

In terms of physical damage the surveying officer is required to look for broken and worn wires. Again there is no authoritative standards for guidance so that recourse must be made to other Authorities of which two are quoted hereafter:—

1. The Statutory Rules and Orders under the Docks Regulations' Section VII, Clause 20c which provides:—
    'No wire rope shall be used in hoisting and lowering if in any length of eight diameters the total number of visible broken wires exceeds 10 per cent of the total number of wires in the rope, or the rope shows signs of excessive wear, corrosion or other defect, which in the opinion of the inspecting person renders it unfit for use.'

2. The 1945 Draft Revision of the Building Regulations, Part III, Clause 58/2, which provides:—
    'No wire rope shall be used in raising or lowering or as a means of suspension if in any length of ten diameters the total number of visible broken wires exceeds 5 per cent of the total number of wires in the rope.'

In order to reduce these two requirements to a single common factor for the purposes, these length of one lay, viz an axial length equal to one complete revolution of one strand of the rope . . . has been chosen as a suitable reference seeing that it can be ascertained without measurement. This gives:–

1. One broken wire for every twelve in the rope in one complete lay.
2. One wire for every thirty in the rope in one complete lay.

Bearing in mind the extraordinary stresses which can present whilst towing at sea it would appear prudent to accept the more stringent Building Regulations as a guide, but because wear on towing wires rarely presents as an even reduction, all about the circumference, the surveying officer must use his own discretion as to the assessment of wear on separate wires as contributory to the loss of whole wires. (See appropriate table in the Appendix).

Deterioration through corrosion in steel wires is possibly the most general cause of condemnation if not for actual breakdown. It consists of two entirely separate conditions. External corrosion due to exposure and abrasion and internal corrosion caused by the loss of the lubricant cum preservative impregnant in the fibre heart.

The external corrosion is readily recognised and relatively easily treated, but the internal corrosion is the more serious because it is invariably not appreciated until it is too late. The fibre hearts in wire ropes are saturated with patent lubricant cum preservative when new but the compression effect of stress tends to squeeze this out and it is eventually replaced by salt water, a condition which is progressive. Good maintenance can delay the onset of internal corrosion but clearly cannot prevent it. This inspection clearly goes hand in hand with examination for stress effect.

As with fibre springs, all of the foregoing effect is cumulative, and must be considered in conjunction with the history of the wire. Here, however, attention must be drawn to the difference obtaining in stress effect in towing pendants and in winch wires. In the former, a long history of towing will only have produced tensile stress which is not ultra productive of fatigue, whereas the same history with a winch wire where there is a very great deal of bending stress at the winch drum and over towing arches and the taffrail, and where vibration is suffered to a secured end, and where crushing stress is obvious, such as at the taffrail and between the winch drum proper and the superimposed coils, fatigue must persist in high degree.

A discussion upon steel wire ropes for towing cannot be concluded without making reference to the physical danger to personnel which is always present when such material is suffering high loading. This should always be the final consideration in the mind of the surveying officer in any doubtful issue, over and above the hazard, inconvenience and expense involved in a spoilt towing operation.

3. The Survey of Nylon and Other Man-made Filament Tow-Ropes.

Because, in this type of rope, there is no deterioration effect resulting from rot, decay, mildew or other bacteria effect, the surveying officers' task is simplified in that his assessments are reduced to the effects of deterioration on two counts:—

1. Deterioration resulting from actual physical or mechanical damage.
2. Deterioration deriving from age and use.

To facilitate the survey the rope must, as with the other types, be fully exposed for inspection and facilities must be available for opening up the rope for an inspection of the interior surfaces.

In the case of (1) above, the surveying officer will examine for broken or abraded filaments on the outside of the rope resulting from chafe against the fittings of the tug or her tows or from abrasion effect from the sea bed. In the case where sea-bed abrasion shows it will be prudent to open up the rope to see whether the rope has not suffered damage through sand, shingle, shells, ash or other sea-bed debris being forced in through the interstices of the rope. Throughout this aspect of the survey the evidence of damage should be related to a length equal to one complete lay of a strand.

Because the filaments in man-made rope are truly homogeneous in construction and are continuous in length throughout the tow-rope regardless of its length, and because they are of uniform circular section and diameter. it is able to accept more physical damage than fibre rope of the same dimensions, but because so much more is expected of this type of rope it is nevertheless prudent to retain the same survey standards as for fibre, viz. to reject the rope for general service if 25 per cent or more of the filaments in one complete lay of one strand are sunken, damaged or defective through abrasion. Because of the miniscule dimensions of man-made fibre rope filaments . . . (See appropriate table in the Appendix) . . . it is not practicable to carry out a specific count of defective units as with fibre ropes so that ordinary practice is to relate damage to the total sectional area of the strand.

Fatigue deterioration due to age and use is best determined by practical measurement and in this aspect of survey the procedure recommended in British Standards Publications, B.S. 3758, 1964 and B.S. 3912, 1965, para. D 2(e), in each case. The Institute recommends that new ropes should be indelibly marked so as to provide reference lengths at regular intervals throughout its length. Surveying procedure then takes the form of a check on these to establish local permanent elongations which, it is considered, offer the best evidence of imminent breakdown.

As with survey procedures with wire and fibre ropes, evidence of deterioration under one count must be associated with evidence produced at another in reaching a decision. In all cases, however, serious doubts as to a rope's worthiness should be sufficient as to merit its withdrawal from service.

CHAPTER V.

The Length and Composition of the Towing Medium.

In the case of towage operations of modest duration, under circumstances of reasonable weather, a towing medium comprising a combination of towing winch wire, in the terms suggested by Section III, Chapter IV, and the assisted vessel's anchor cable is quite satisfactory; the strength of the gear will be competent to the occasion and the flexibility provided by the towing winch will satisfy the demands of an uncomplicated operation. When longer tows are envisaged where substantial ocean passaging forms part of the operation, this arrangement is no longer adequate and it becomes necessary to introduce some form of auxiliary shock absorbing material into the towing medium for two reasons:—

1. To ensure some degree of emergency elasticity in the towing medium in the event of whole or partial loss of the towing winch facility.

2. To allow for the less apparent, but wholly feasible circumstance that the stress of towing plus the weight and resistance to forward movement offered by the towing medium, together with such added resistance as may derive from protracted adverse weather or a physical deterioration to the vessel, or object, in tow might be such as to exceed the upper compensatory limits of the towing winch so that it must be spragged.

Whilst the case of 1. is always readily appreciated, that at 2. is not so, so that the following hypothetical case is offered in amplification.

Consider the towage of a freighting vessel of about 8,500 tons gross, 5,000 tons nett and of a displacement tonnage approximating to 17,000; possibly about 475 ft. in length and with a beam of nearly 60 ft. and loaded to a draught of $27\frac{1}{2}$ ft. The tug involved being of the Class 3 Sea Going Tug of about 2,500 B.H.P. using about 305 fathoms of gear made up of 45 fathoms of the ship's $2\frac{1}{2}$ inch anchor cable, 60 fathoms of 18 inch Manila or double 7 inch nylon and perhaps 200 fathoms of $5\frac{1}{4}$ inch E.S.F.S.W. towing wire.

It would be practicable to assume that in this case two out of the three shackles of chain would be wholly immersed as would be the whole of the spring and **180** fathoms of the wire.   The actual suspended weight of this combination would be:—

15 fathoms of cable in the air	5155 lbs.
30 fathoms of cable in the sea	9428 lbs.
60 fathoms of manila or nylon spring, mostly fittings	250 lbs.
180 fathoms of wire in the sea	4401 lbs.
15 fathoms of wire in the air	424 lbs.

A total of 8·8 tons, supported by both tug and tow, so that the towing winch must hold one half of this weight before transmitting any of the towing strain.

A little elementary arithmetic will show that the immersed parts of the towing medium offer a wetted surface of approximately **1,200** square feet and that the dimensions of the vessel towed (in this hypothetical case) suggest a wetted surface of about **45,000** square feet so that, in the terms of pure resistance to forward movement, the towing medium must provide appreciable additional burden.   The wetted surface aspect is not, of course, the whole of the matter seeing that the towing medium is never quite still in relation to either tug or tow but is in almost continuous movement as it adjusts to the lateral movement of the tow in terms of yaw, or longitudinally in the case of surge.   The medium also hangs in a catenary of quite considerable proportions most of the time so that, besides the attitude and lead of the medium to the line of advance, there are complications arising from pressure at the not inconsiderable depths involved.   Finally, there is the extra resistance to forward movement offered by the bulk of material at the connections of the separate parts of the towing combination, particularly when fibre springs are in use.   All of this has been most exhaustively examined by the United States Navy at their Experimental Tank at Washington, D.C., when it was established that tow-line resistance could amount to the rather surprising figure of **13** per cent of the ship resistance.

If the tug used in the hypothetical consideration was fitted with an automatic winch with an upper compensatory limit of **30** tons and the smooth water true pull required to tow the vessel at a reasonable navigating speed was say **10** tons, then the weight of the gear in use together with the resistance of its immersed parts to forward movement, might well augment this by **5** tons or more so that the actual tension at the winch would be in excess of one half of the capacity of the winch so that less than **15** tons of compensatory force would be available to cover the important stresses occasioned by wind and sea, and the movements of both tug and tow resulting therefrom.

It is then most seriously submitted that whilst the shorter towing operations may well proceed using a simple wire and cable combination towing medium, seamanlike prudence clearly demands the inclusion of some emergency provisions to ensure elasticity when longer ocean tows are contemplated.

When it can be demonstrated that up to 13 per cent of towing stress can derive from the medium in use it becomes apparent that, whenever circumstances permit, the speed of a towing unit can be increased if the tow is shortened. Indeed, in calm weather, when there is no substantial movement of tug or tow, full advantage must accrue from lifting the gear clear of the water. In such case the automatic winch compensatory gear will give ample warning of deteriorating conditions.

The prime guide in terms of the total length of towing medium to use under any combination of practical conditions must be that which allows of the best forward progress yet permits of none of the shock stresses to the towing medium, 'in part or whole,' which are the prime cause of tow-rope failure.

When vessels or objects towed are of a substantial draught, the tug's screw race has a markedly decelerative effect upon the progress of the tow, particularly when under-keel clearances are small and when the frontal aspect of the tow offers a full or flat aspect.

Experience deriving from the towage of some of the very large objects required for the invasion of Normandy in 1944, and for the subsequent maintenance of the armed forces ashore showed that this effect, in the case of single tug working, could only be reduced by veering a sufficient scope of towing medium to allow of some dissipation of the screw race. In the case of tows having draughts of from 24 ft. to 30 ft., in depths which permitted of only a few feet of under-keel clearance, this ill-effect could only be eradicated by veering upwards of 1,000 ft. of towing gear.

All of the foregoing applies, of course, to tugs fitted with automatic towing winches. There are however, quite a number of ocean tugs in service which are not and these will, as standard practice, always include a fibre or a nylon spring into ocean towing media. The length of spring used will vary as between tugs and according to the operation, but whereas, in the case of winch fitted tugs, the spring provides emergency utility only, in non-fitted tugs the spring is regarded as the prime component of any towing combination. Fibre and man-made filament springs are normally turned out by the manufacturer in lengths of 120 fathoms and most tugs employ this entire length as a spring in any ocean-towing operation of any consequence, together, of course, with steel wire rope pennants of a total length appropriate to the occasion.

Although adjustments to the length of a towing combination are very difficult of achievement in tugs where no automatic winch is fitted, this operation must occasionally be effected in the interests of the safety of an operation. It is always easier to do this when towing is done from a towing bollard rather than from a towing hook but, in both cases, the process is much facilitated when the towing spring is suspended, as it were, between two steel wire rope pennants. This practice also provides the important side effect of reducing chafe damage.

In addition to the remarks offered in the matter of towing equipment in Section III, attention must be drawn to the necessity of using cable joining type shackles when making up towing sets, for all of the reasons already offered. It is also a clear practical arrangement to fit the shackles so that their bows face forward in the set.

CHAPTER VI.

Tow-rope Handling.

Such complications as do arise in the handling of towing gear ordinarily only derive from their greater than average bulk and weight. In example of this it may be quoted that an average type towing spring, in one of the fibres to a length of one hundred and twenty fathoms and of twenty inches circumference, fitted with thimbles and links, will weigh upwards of three and a half tons and will present itself in a close manufacturer's coil measuring approximately seven feet by five feet. A new winch wire, for the same class of tug, of a length of three hundred and fifty fathoms and to a circumference of five and a half inches, will arrive on a manufacturer's wooden reel about seven feet in diameter, the whole weighing in excess of five tons. The practical aspects of tow-rope handling are, however, possibly best described under the following three broad cases:—

1. Boarding new gear.
2. Operational handling.
3. The protection and preservation of tow-ropes whilst they are in use.

1. Boarding New Gear.

Towing springs are supplied alongside of the tug in the manufacturer's close coils, as with all other cordage, except that towing springs are commonly supplied ready provided with a thimble and link already spliced into each end. When space and time allow it is the common tug practice to take springs out of the manufacturer's coils and to stow them on board of the tug in ready-to-use loose coils immediately upon delivery.

The manufacturer's coil is set up on its side so that the finishing fleet face of the coil is disposed at right angles to the direction of the first lead or bearing point on board of the tug. A messenger is then brought off to the coil from the tug via the leads required for the haul off and is rove through the centre of the coil and is made up to the inner end of the coil. This end is then carefully eased through the centre of the coil by taking the messenger end to the capstan and heaving slowly. Once the thimble and link are clear, the inner lashings of the coil are cut and heaving out can proceed. When sufficient end is brought inboard of the tug, turns of towing spring may be placed directly on to the capstan to advantage.

The principal problems encountered in carrying out this very simple and straightforward operation are those presented by the need to keep the coil together and firmly in place to withstand the hauling off strain. A pair of fork lift trucks provide the ideal supporting medium for this proceeding but when these are not available recourse must be made to any heavy supporting material as lays to hand such as barrels, cases and the like. This essentiality always taxes the resource and ingenuity of tug personnel, particularly when new gear is supplied at remote places, but care must be taken because a collapsed coil not only presents a most embarrassing spectacle but is productive of a deal of quite unrewarding toil and trouble.

Many tugs elect to provide themselves with portable turn-tables made up of members designed to unite to provide an 'X'-shaped device of dimensions and strength suitable to the tow-rope establishment. Such a device is ordinarily provided with a flange based swivelled eye at the junction of the members. A coil for re-fleeting is lowered on to this device and then the derrick fall is rove through the centre of the coil to engage into the swivel eye and the coil lifted a few inches clear of the deck. Refinements to this device consist of lids, tops or other arrangements designed to prevent turns raising off the coil to foul proceedings and arrangements of stays to hold the unit steady. However, with the coil ready for treatment, the outer lashings are cut and the spring is taken off the coil from the outside by revolving the turntable.

To assist with the reception of new winch wires certain tugs have arrangements worked into basic fabric. These normally consist of a suitable bracket, arranged in partly portable form, made up to the bulwarking at a convenient location with a wholly, or partly portable pedestal situated at a suitable distance from it. These two fittings provide trunnions for an axle to support the towing wire reel at an aspect proper to hauling off. When these arrangements do not obtain, and the derrick gear can accommodate the weights involved, a hanging turntable may be utilised, failing these, recourse must be made to a pair of jacks in conjunction with an axle.

It is more often than not the case that new winch wires are required to be rove on to the towing winch immediately upon delivery and this operation presents complications which are not readily apparent because of the necessity of reeving the winch wire on to the winch barrel as tightly as may be contrived. This precaution is quite essential if, under the subsequent stresses of towage, the last turn on the barrel is not to pull down though its supporting layers to set up complications, the least of which are the physical distortion and lacerations which are inflicted upon the wire.

To this end it is the common tug practice to heave winch wires on to the barrel against an element of quite considerable frictional resistance and to settle each turn, and each layer of turns, soundly and firmly against the drum flanges, and the other turns and layers, by the use of a large wooden mallet. The frictional resistance to heaving to impart tightness to the turns is obtained by reeving the winch wire back and forth across the deck through old towing shackles set up to substantial ship's fittings. It is not permissible to obtain the desired resistance by taking turns about the tug's bitts or around shore bollards because of the deep scoring which results.

The processes at the winch consist of operating the spooling carriage clutch and hand-wheel adjuster so as to align the spooling carriage rollers precisely in line with the appropriate drum flange. The whipped, tapered or ferruled end . . . (according to design detail) . . . is then passed through the spooling carriage rollers and is rove through the orifice at the junction of the winch drum and flange to be secured by the clamp fitting disposed on the outer side of the flange. The wire is then set up taut and the alignment rechecked before the spooling gear is clutched in prior to commencing the heave on. It is the common practice to apply a suitable form of lubricant plus preservative to the wire, usually at the location between the drum and the spooling gear, to replace that which will be removed by the heaving on processes.

Shorter lengths of towing wire, in the form of pennants, are usually rolled out of their coils in the time honoured manner. Lengths too great for this treatment are customarily referred to the turntable.

It is perhaps permissible to observe that the processes of reeving new winch wires are laborious enough when a tug is lying in port and alongside of a wharf. At sea, particularly if the weather is, in the slightest degree inclement, these processes become difficult to the point of near impracticability. Liaison between the seamen and the winch manufacturer is clearly imperfect seeing that these difficulties are not resolved when they are, so clearly, reasonably easy to resolve.

2. Operational Handling.

(a) Veering or Streaming the Towing Media.

Fibre or nylon springs are veered or streamed from a loose coil or flake disposed conveniently on deck. There is not a deal of remark to be made concerning the coiling or flaking down except perhaps to observe that a clear run is quite essential so that successive fleets in the coil or flake must be laid so that the upper fleets lay fair and clear over those beneath.

Control over these ropes whilst fleeting is customarily imposed by means of a substantial stopper rigged in the ring fashion. Suitable ring anchorages for stoppers form a part of most tugs' initial equipment and are usually arranged on each side of the towing deck at convenient afterly locations.

The streaming and veering of steel wire ropes are much more precise affairs where the winch mechanisms, with their automatic braking and overloading devices, ensure that full control may be imposed with safety. Winches normally stream and veer with the winch in the pay out condition so as to prevent any possible chance of the gear taking charge, although it is permissible to veer or stream with the winch out of gear, control being exerted through the manually applied hand brake when circumstances allow.

(b) Tow-Rope Control Inboard of the Tug Whilst Towing.

Strict and full control of those sections of tow-ropes which lie inboard is quite essential whenever towing proceeds. This for two reasons. The first is so that the movement of the tow-rope in arc may be restricted to the minimum so as to reduce chafe effect. The second is to ensure that the tow-rope shall not sag down close alongside, through the movement of the tug and the tow during adverse weather, to introduce a risk of fouling the propeller. When working at close quarters or in company with other tugs firm control must be maintained to reduce the risk of Girding . . . (Of which more later) . . . Close quarters working especially requires both control and vigilance so as to assist the tug in her manoeuvres and so as to safeguard her should sternway be obliged upon her by the manoeuvrings of the vessel which is being served, either as a result of other tugs' movements or because of movement induced by main engines or other causes.

The desired control is exerted through the application of a rope which is commonly known as a bridle but which is also known as a "Gog-Rope."

or a "Gob-Rope." Control is obtained by securing one end of this rope to a secure anchorage known as a "Gogging" or "Gobbing" bollard, passing it through a lead known as a "Gog" or "Gob" eye, thence over the tow-rope and back through the 'Gog' or 'Gob' eye to make fast once more on the 'Gogging' or 'Gobbing Bollard'. This rig is known as 'Bridling on the Bight' and is always practiced when applying bridles to fibre or nylon towing springs. When this gear is rigged to a wire tow-rope the more usual procedure is to pass a bow shackle over the tow-rope and to make up the bridle to this.

When towing at sea, the bridle is always made up from the same material as the main towing medium, thus; a Nylon rope would be bridled with nylon, fibre with fibre and steel wire with steel wire and quite substantial stuff must be used for the purpose in view of the punishment which must be suffered, particularly with yawing tows in heavy weather.

When more than one vessel, or other floating object, is towed simultaneously, it is very important to see that the tow-ropes to the separate units in tow do not foul up inboard or at the taff-rail because of the extremely complex stresses which result. Most modern tugs are provided with 'Gogging Eyes' and 'Gogging Bollards' which are expressly designed to accommodate up to three bridles simultaneously.

Towing at short stay requires both vigilance and skill in the manipulation of the bridle when large tugs are employed. When this type of towage is engaged upon a relatively light bridle is employed which is set up so as to bring the working end back to the capstan barrel instead of making fast. The scope of bridle can then be veered and recovered in sharp time in concert with the tug's movements upon the rope. Besides the control exerted in keeping the slack of the rope up clear of the tug's propeller, and the securing function desired to prevent girding, a skilled hand with a bridle can actually assist the tug's helm in swinging the tug by exerting weight whilst effecting the recovery.

(c) Handling Gear Whilst Shortening In.

In the case of the recovery of steel wire rope using a towing winch, there is little to add to the detail in this offered in the description of tugs' manoeuvring whilst thus engaged. One observation which is possibly permissible is that of emphasising the necessity of re-greasing the wire as it is recovered, particularly if it has been in contact with the sea bed.

Fibre gear benefits from a vigorous hosing as it comes in over the rail, particularly if the shortening in processes have drawn the gear along the sea bed. Besides the sand, shingle and sea shell which are forced into the interstices of the rope when this happens, it must be remembered that the sea bottom in the approaches to most large seaports is fairly well fouled up with debris of one kind and another which might well inflict serious damage if it is not seen and removed. The risk of 'Gilding the Lily' is accepted in offering the recommendation that the hosing may, to advantage, be carried out with fresh water.

It is no part of this work's function to repeat the perfectly satisfactory descriptions of stoppering which appear in all of the standard seamanship works, it is however the truth that 'a very great deal hangs upon a stopper' particularly when dealing with towing media. It is therefore incumbent upon Tug Officers to make sure that all ratings can apply a stopper properly to either fibre, man-made fibre, or steel wire ropes.

3. The Protection and Preservation of Towing Media Whilst in Use.

All towing media suffer chafe damage whilst towing, nowhere more seriously than at those sections which lie inboard of the tug. In the case of steel wire ropes the chafe over the towing arches need not be too acute if the arches have been designed and fitted with a sympathetic understanding of the problem. In such cases the standard practice is to anoint the arches liberally with a stiff mineral grease and to ensure, by repeated and regular inspection, that a satisfactory film of lubricant is permanently offered to the wire at every bearing position.

Chafe at the tug's aftermost bulwark railing is a much more serious problem because, regardless of the naval architect's skill, the angle of dip of the tow-rope cannot be materially changed nor can a wholly friction-free surface be devised. Again the critical area may be generously lubricated but the only practicable paliative resides in spreading the load by making frequent adjustments to the scope of gear in use. This proceeding effectively substitutes wide-spread minor effect for acute localised damage but clearly the overall damage suffered is not thereby reduced. In their search for a means of reducing chafe effects tug personnel have improvised a variety of devices of which only one has endured. This is an appliance which is known either as a 'Dutchman' or a 'Scotsman' and which consists of a length of tubing, usually about four feet in length and of a diameter somewhat in excess of the wire to which it is to be applied. This tubing is split longitudin- ally and provided with bolts and clamps for securing it to the wire. The

I

two halves of the device are lined with soft-wood, plastic material or canvas, and the more carefully made examples have flared ends. Although these devices are not easily produced they are of limited value because they tend to promote a strip of rigidity in the otherwise flexible whole of the threatened section of wire with the obvious attendant ill-effects at the ends. They are moreover only effective as long as the bridling can be arranged to restrict the movement of the towing gear, in arc, about the afterparts so as not to exceed the length of the 'Scotsman'. It will also be apparent that this sort of gear cannot be used if the automatic compensating winch effect is needed. (Figure No. 32).

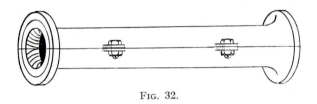

Fig. 32.

The whole answer to this chafe problem, in its application to steel wire ropes, clearly only obtains with the provision of a flexible steel sleeve device which can be maintained over the taff-rail by means of some device which resolves the movement of the towing medium in arc into lineal adjustment along the towing medium.

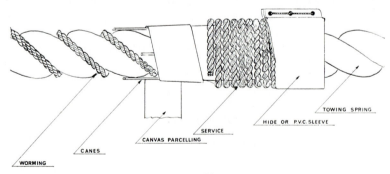

TOWING SPRING

HIDE OR P.V.C. SLEEVE

SERVICE

CANVAS PARCELLING

CANES

WORMING

Fig. 33.

Towage to a fixed steel wire end is not the best towing practice but it is still infinitely preferable to towing to a fixed fibre or nylon end inboard. This is a situation which must be avoided whenever possible because of the chafe damage which results. Damage from this cause over the towing arches can be contained by rigging hide or heavy PVC sleeves over the rope bearing upon the arches and then applying generous coatings of animal or vegetable based lubricants thereto. Care must be taken to see that the lubricant is kept clear of the rope of course because of the ill-effect which results. Taff-rail chafe cannot be so conveniently contained for the same reasons as were offered above and many and varied are the efforts made by tug personnel to reduce its effect. The general aim of such efforts is directed towards the provision of protective material which will be sufficiently flexible as to conform to the changing aspect, caused by the movements of the tow, over the troublesome areas but which will not harm the medium either because of its own strength or in the processes of breaking up in service. The common practice is to worm, parcel and serve the rope, over the critical length, with old rope and canvas finishing up with an outer cover of tanned hides. Some officers of ingenuity have found means of working flexible canes into such arrangements but they are all somewhat short lived in all but the most clement of weather and therefore require constant attention, besides which the actual labour of maintaining and repairing such arrangements, particularly under bad weather conditions, is both heavy and unrewarding. (Figure No. 33).

CHAPTER VII.

Sag in Ocean Towing Media.

The sag which develops in open seas towing media has, so far, only been touched upon in terms of the margin of elasticity which is thereby provided and no comment has yet been made concerning the actual extent to which this may develop in the course of practical usages. This aspect of the phenomenon is, however, of the most critical importance to the safety of tugs and their charges when they are on passage through shallow waters; most particularly when areas are traversed where the debris of both warlike and commercial intent and accident litters the sea bed.

Close observation of the attitudes assumed by towing media over the after parts of ocean-going tugs, whilst towing to long scopes of towing gear, reveals that the angles of dip in ropes are not constant, even when vessels or other floating objects in tow follow the tug quietly and at apparent good station. Checks provided by a series of vertical sextant angles or radar ranges would appear to show that, even under the most favourable conditions, there is an element of surge usually present. Surge effect seems to show at its greatest when large vessels of ship-shaped form are in tow and reduces to a minimum when other than ship-shaped objects, offering a higher resistance to forward motion, provide the burden of ocean towage operations.

At its most evident surge effect would appear to indicate that open seas towage rarely, if indeed ever, presents in the form of a smooth and continuous application of effort. Rather would it appear to consist of a succession of separate phases begun when a tug first moves forward through the water to take up the greater part of the sag in the tow-rope so that forward motion is not imparted to the object towed until a particular moment is reached when the tug's way through the water diminishes whilst tow-rope tension increases sufficiently to overcome the tow's resistance to forward motion. When this condition is achieved, the tow gathers forward momentum to allow the sag in the towing medium to reform thus allowing the tug to move forward once again to initiate another phase in the sequence.

Such observations suggest an acceptance of the fact that any degree of sag in towing media can rarely maintain at a constant depth even under the quietest of operational conditions. When the circumstances of wind and weather become adverse then the separate motions of both the tug and her charge must clearly impose additional effect, a circumstance which must be still further complicated should yaw also be experienced. (Fig. No. 34).

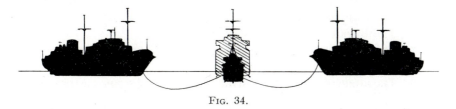

FIG. 34.

Another qualification to sag effect in towing media is provided by the character of the towing medium itself. If the gear in use comprises one continuous length of steel wire rope of uniform diameter it should, if the tow could be halted momentarily in the water, hang in a bight between the tug and her tow of true catenary form, but if the medium is a composite one having elements of steel wire rope, manila fibre or nylon, together with an amount of anchor chain cable, then the sag form will be of unsymmetrical form according to the disposition of the elements. (Fig. No. 35).

FIG. 35.

In any case it must be assumed that the forward motion of a towage unit through the water will also impose additional distortional effect.

It must, therefore, be accepted that no precise calculation of sag values is ordinarily practicable for seamen afloat. Nevertheless, it is quite imperative that means be provided for the estimation, even in approximation, of the extent to which this phenomenom will develop under ordinary conditions of towage.

In seeking to provide such an estimation the following data is reasonably accessible in most ocean-going tugs:—

1. The precise length of towing medium in suspension between the aftermost parts of the tug and the first bearing point inboard of the tow.
2. The precise distance between the two points mentioned at 1, above.
3. The composition and weight of the towing medium in suspension.
4. The towage effort being exerted by the tug.
5. The respective heights above w/l of the two points mentioned at 1, above.
6. The angle of dip of the towing medium at the tug's after extremity.

The detail required at 1. above is readily obtained in all tugs. The distance required at 2. is conveniently obtained by sextant angle or radar. The weight of gear in use is available from the Appendix to this book, allowance being made for immersion of course. The requirement at 5. is a matter of organisation, and that required at 6. is readily achieved by any tug seaman of initiative. This leaves only the important requirement indicated at 4. above.

An accurate measurement of the towage effort being exerted will, in all probability, provide the major concern in the suggested estimations. When the only instrument competent to this occasion was a dynamometer . . . (an instrument which, regardless of its precise shape and form, presents expenditure in excess of prudent usage) . . . there was possibly some excuse for not providing every tug with the means of determining tow-rope tensions, but, in more recent years, the Micro-Strain Gauge as developed by H. Hitchens, Esq., of British Ropes' Technical Service Department, has become available to provide the desired information efficiently and economically also, if it is so desired, in the form of a permanent record. (Fig. No. 36).

The Micro-Strain Gauge is a very compact and sturdily constructed extensiometer measuring 31 in. × 9 in. and weighing 157 lbs. nett. It is provided with a coupling lug at each end so that it may be shackled in to a convenient anchorage and set up to the tow-rope by means of a carpenter's stopper. The body of the device is a one piece forging machined so as to provide three equally spaced tension bars joining the coupling lugs. The material used for this is steel of a tensile 105 tons/in. with a breaking strength of 315 tons.

Located into the space between the tension bars resides a recording device activated by a small electric motor powered by a pair of dry cells. This equipment is competent to continuous experiments of up to thirty minutes duration but for tests of greater duration a mains supply may be arranged as an alternative to an appropriate accumulator equipment. The motor serves to traverse a length of micro-film over a suitable table utilising a pair of reels of appropriate form and size so that the extension of the bars under the stress of towage is recorded thereon by a properly located stylus.

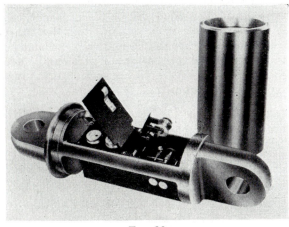

Fig. 36.

Tensiometer or Micro-Strain Gauge.

Permission of British Ropes, Ltd.

Without the assistance of such a device, or one of its more complicated and expensive predecessors, towage effort may only be estimated, to the obvious detriment of the exercise, although it is, at once, conceded that an intelligent use of Towing Winch performance diagrams can ameliorate this deficiency somewhat.

The angle of dip over the tug's after parts cannot presently be utilised to assist in the estimations envisaged but it is offered that all forms of towage observations should be as comprehensive as circumstances will allow, even if only for future reference in the case of similar operations.

With the availability of this data a very reasonable estimation of sag values becomes a practicable proposition. There are a number of formulae available which will provide precise catenary proportions under the ideal circumstances of suspension, but these are not applicable in the case of towing media for two reasons. The first is that practical circumstances, for reasons already offered, do not encourage a purely mathematical approach to the problem and the second is that Ocean Tug Officers require the desired values quickly and conveniently, usually under circumstances when there will neither be the desire nor the facilities to enter upon the resolutions of complicated formulae, particularly if they involve the use of hyperbolic values of expressions or adventures into the Calculus. It is, therefore, submitted that the simplest method will be the best method.

With this in mind it is seriously offered that the technical differences between catenaries and parabolae may be ignored especially bearing in mind the approximate nature of the estimations. If this can be accepted then the following transposition of terms in a formula, normally associated with suspension bridge cables, is submitted:—

$$b = \frac{wa^2}{4\sqrt{4T^2 - w^2a^2}}$$

Where:—

a = The precise distance as instanced at 1. above.

b = The depth of the sag in the towing medium.

w = The loading per foot of horizontal run. Viz. the weight of the towing medium as instanced at **3**, above (allowing for the degree of immersion) divided by the value a.

T = Tow-rope tension measured in pounds.

Consider the application in the case of an ocean-going tug towing a merchant ship using a towing medium comprising 2,000′ of 1¾″ E.S.F.S.W.

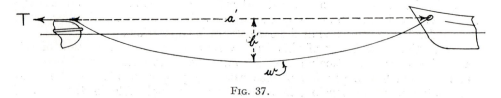

FIG. 37.

(Diameter) 90 per cent immersed. Tensiometer reading of 8 tons at the moment that vertical angle observations show the a value to be 1960 feet.

In establishing the value of 'w':—

90% of 2000′ of $1\frac{3}{4}''$ E.S.F.S.W., immersed $=1800 \times 3 \cdot 87 = 6966$ lbs.

10% of 2000′ of $1\frac{3}{4}''$ E.S.F.S.W., in air $\quad = 200 \times 4 \cdot 93 = 986$ lbs.

$$w = \frac{6966 + 986}{1960} = 4 \cdot 05$$

So that:—

$$b = \frac{4 \cdot 05 \times 1960^2}{4\sqrt{(4 \times 17920^2) - (4 \cdot 05^2 \times 1960^2)}}$$

$$b = \frac{15558000}{4\sqrt{(1284500000 - 63015000)}}$$

$$b = \frac{15558000}{139800}$$

$$b = 111'$$

But this depth is measured from the bearing points on board of the tug and tow (5. above). If, in the case of the tug, this was 10′ above $w/1$ and on the tow 40′ above $w/1$, the mean height would have been 25′ which must be subtracted from the calculated 'b':—

$$111' - 25' = 86'$$

This being the depth of the sag in the towing medium under the conditions described.

For comparison consider the case of a tug applying 35 tons of tow-rope tension to a very heavy floating object using 2000′ of 2″ E.S.F.S.W. which is 90 per cent immersed. The value of 'a' being, in this case 1990′.

In establishing the value of 'w':—

90% of 2000′ of 2″ E.S.F.S.W., immersed $=1800 \times 4 \cdot 92 = 8856$ lbs.

10% of 2000′ of 2″ E.S.F.S.W., in air $\quad = 200 \times 6 \cdot 32 = 1264$ lbs.

$$w = \frac{8856 + 1264}{1990} = 5 \cdot 06$$

So that:—

$$b = \frac{5 \cdot 06 \times 1990^2}{4\sqrt{(4 \times 78400^2) - (5 \cdot 06^2 \times 1990^2)}}$$

$$b = \frac{20038000}{4\sqrt{(24588000000 - 101390000)}}$$

$$b = \frac{20038000}{62588}$$

$$b = 32'$$

Again, subtracting the mean height of the bearing points which, in this case might well be 11′, then:—

$$32' - 11' = 21'$$

would be the maximum degree of sag below w/l under the circumstances described.

A practical seaman's reaction to the foregoing submission might easily be that the suggested procedure is somewhat involved for the estimation of a factor which must, by the very nature of the exercise, provide an approximation only; suggesting the corollary that more elementary methods could well provide results which are only marginally more approximate. To satisfy such considerations the following simple procedure is offered.

Knowing the length of gear in use and the distance between the tug and tow; cut a piece of waxed thread to a suitable scale to represent the former and apply it, by the ends at the proper distance apart, upon a horizontally disposed scale to represent the latter. The bight then provided by the excess of the former over the latter, measured to the same scale, will approximate to the degree of sag obtaining under the circumstances of the observations.

SECTION 4.

CHAPTER I.

Practical Ocean Towage Seamanship.

At the outset of this discussion attention is drawn to the importance of appreciating the fact that vessels of different types and sizes, in all of the infinitely varying conditions of lading, ballastage and trim occasioned by their service and trading, take up widely differing angles of rest in the sea when deprived of main propulsion power. These angles of rest relative to the wind and sea depend upon a number of factors, the most important of which are:—

1. The amount and disposition of windage offered in terms of super-structure and freeboard.
2. Resistance offered by the sea water to the form, amount and disposition of the immersed hull.
3. Fore and aft trim.
4. Underwater drag offered by the propellers, rudders or any under-water damage which may provide part of the casualty.

The angle of rest will determine a vessel's motion in a seaway, from pure roll in the case of a vessel lying precisely beam on to the elements, to a straight pitch and scend in the case of one lying truly head on or stern on. More complex situations arising, of course, in the case of vessels lying at broad angles to the weather when the resultant action, commonly referred to as 'Corkscrew Roll,' causes bow and stern to yaw considerably as these extremities encounter sea action before and after the amidships parts.

A vessel's rate of lee drift is clearly conditioned by her angle of rest in the sea seeing that the total wind pressure upon the hull will vary proportion-ately to the aspect offered.

It is apparent that any vessel which lies exactly athwart the wind will sag directly down to leeward, and that vessels lying head on or stern on to the wind will make straight stern-way or head-way respectively, so that it is obvious that the lee drift made by vessels constrained to lie obliquely must be of a composite nature made up of leeward sag plus some measure of head-way or stern-way depending upon whether they are lying head off, or head on to the wind, and to what extent.

All of this is offered at some length because attention must be drawn to the consideration that, except in the very unlikely case of a tug being called upon to lend assistance to a sister ship, she must, quite inevitably, react differently in this respect to practically every vessel which she is called upon to assist. Unless therefore a towage connection is effected under conditions of total calm, complications must arise because not only will the tug and her intended tow drift in differing directions, they will do so at quite different speeds.

Because of the weight of the gear to be passed, the success of all connecting up operations depends upon a wholly safe and well-judged approach followed by the closest possible attendance until the processes of connection are complete. In example of this, a typical towing medium, $5\frac{1}{2}$ in. E.S.F.S.W.R. weighs 4·93 lbs. per foot (44 mm. diameter 7·34 kilos per metre). If tug and prospective tow are 100 ft. apart, the bight of rope suspended between the two may well be twice this distance in length and will weigh close on to 1,000 lbs. Under circumstances of adverse weather, even with power to hand, and this is not always the case with a casualty, such weight can provide its own difficulties.

Holding a large tug a bare 100 ft. away from a disabled ship in open seas conditions is work for an expert so that it follows that average operating conditions will be somewhat greater, it is clear however that any distance in excess of 100 yards will, generally speaking, be impracticable, excepting under the most favourable of circumstances.

In order that this vital aspect of sea towage operations shall be fully understood a rather detailed appreciation of circumstances now follows. For this appreciation a typical ocean-going tug type has been selected, together with certain ship-types whose marked profile characteristics present the best example material.

Angles of Rest and Lee Drift in the Consideration of Effecting a Towage Connection in the Open Sea.

It is perhaps logical to consider the tug herself first in this respect and Figure No. 38 offers a profile presentation of a typical large ocean-going tug of something in the vicinity of 1,000 gross tons. The lee drift and angle of rest characteristics of this type of vessel are conditioned by three major design features, the first is the totally unobstructed towing deck extending over approximately one half of the tug's length, the second is the top-gallant forecastle head which is an essential seaworthiness feature and last, the keel-rake which is required to ensure optimum propeller immersion under all circumstances. (See Section II).

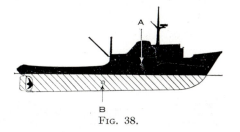

Fig. 38.

These three characteristics have the effect of causing a tug when stopped in the open sea, under conditions providing wind strength of any consequence at all, to take up a stern to wind attitude. This because the wind force, acting about the centre of gravity of the exposed freeboard and superstructure at *A* in figure 1, and the sea water drag exerting more effect abaft of mid-line than forward, as indicated by the focus *B*, assisted by the additional drag offered by rudder and propeller, both proportionately very large in tugs, form a couple of sufficient strength to hold the tug stern up wind. The stronger the wind the more truly will a tug lie.

Although the superstructure and freeboard windage offered by the tug in this attitude will be the minimum achievable, this condition must be considered with the fact that the tug will be wind-driven head-on, in the easiest direction. Lee drift in tugs is therefore head-on, down wind and at an appreciable rate.

Because it is of the greatest practical importance, another effect of this attitude to wind must be mentioned at this early stage. When winds are of about Force 4 and upwards, most ocean-going tugs do not exhibit the normal transverse propeller thrust effect when the engines are put astern, instead they point their sterns up into the wind and run astern dead to windward. This well defined handling peculiarity, intelligently anticipated and properly utilised, can react to the greatest effect whilst effecting towage connections at sea.

Figure No. 39 represents a selection of typical ship profiles and probable angles of rest which such types may well assume under the effect of consequential wind force whilst stopped in the open sea; the sketches also endeavour to illustrate the probable direction of lee drift in each case. No apology is offered for any possible over-simplification presented by these figures seeing that precise angles of rest and directions of drift will be conditioned by so many ancillary factors that a detailed review of all of them would require a discussion over-long and too involved for the practical purposes of demonstration. If the figures convey the broad principles involved then their purpose is achieved.

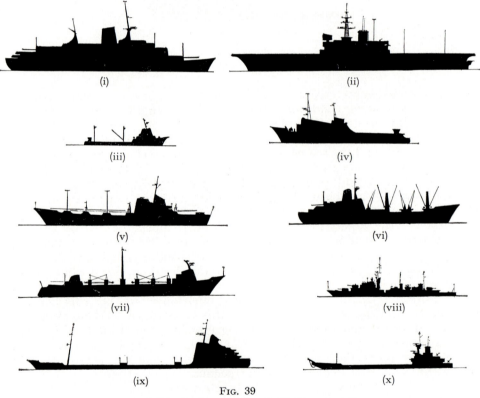

FIG. 39

Typical Ship Profiles and Lee Drift Characteristics.

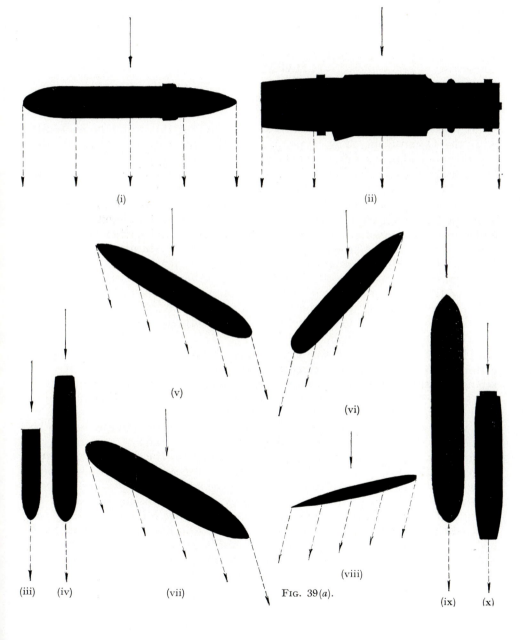

(i)

(ii)

(v)

(vi)

(iii) (iv)

(vii)

(viii)

Fig. 39(a).

(ix) (x)

CHAPTER II.

Approach Manoeuvres.

It must be emphasised that this section, indeed this whole book, treats with the Ocean-Going Tug Type particularly, the assessments made and the recommendations offered being referred to this type only and not to the general run of Merchant Vessel or the various Warship types treating with towage as an isolated emergency measure, or as an evolution.

The descriptive matter hereafter submitted derives from methods developed through experience on practical operations tempered by advice and information emanating from a number of interested parties, and conditioned by the closest study of operational reports made available by Tug Personnel and Operators associated with tugs under a variety of service conditions. The methods to be described however are not claimed as the only, or the proper methods seeing that Ocean Towage is such an intensely individualistic calling that there will always be a number of practical solutions to each and every problem. The methods offered are however both safe and practicable.

The fully experienced tug Commanding Officer is, of course, able to estimate angles of rest and directions and rates of lee drift without even seeing the ship concerned, providing he is given an accurate accounting of the ship, her condition of lading or ballastage, and the weather obtaining at the time. With a vast store of experience behind him he can appreciate, well in advance, the relative behaviour of his own vessel and the one to be assisted so that he is able, upon arriving upon the scene of a casualty, to conduct his tug to an appropriate position relative to the drifting vessel and there literally await the arrival of the casualty into a desired position where he will have his tug correctly disposed for the expeditious passing of gear. This is most impressive to see but is clearly only possible in the case of an expert. This account is not written for the expert, but is designed as a guide to a newly appointed tug Commanding Officer who may, or may not, have a deal of towing experience behind him, and is intended only to serve as a guide whilst the necessary experience is in process of accruing, and whilst new individualistic techniques are developing.

132

It is therefore initially offered that every prudent tug commander will stop his tug in the close vicinity of any ship to be assisted so as to reaffirm the tug's angle of rest and rate and direction of lee drift, making detail alterations to the trim of his own ship, by the disposition of water ballast, to obtain a completely satisfactory condition. He must also, by the best means to hand, carefully estimate these same factors for the disabled vessel so that the relative movements of the two vessels are established. This should be followed by steaming the tug slowly around the casualty so as to accurately observe her movement in the water in terms of pitch, roll or composite movement.

The tactical manoeuvre of approach and connection should then be planned accordingly. In view however of what has already been offered, when wind force is of any consequence, the tug Commanding Officer is recommended to make a down-wind approach because this is the tug's natural attitude when stopped and this is the attitude, relative to the wind, which she assumes under the effect of astern engine movement. The ocean-going tug type will proceed better down-wind, under complete control, under the slowest possible speed through the water, than in any other direction. It is, in fact, practicable to sail the modern tug type before the wind down to a disabled vessel, under full control, with only the occasional touch of slow speed ahead. In the event of untoward occurrence therefore the tug has only minimal headway and can be rapidly extricated from danger by an astern movement which will remove her dead up-wind. In the event of a slight mis-appreciation of composite drift, such as with a vessel lying obliquely to the wind, the tug may be held, still truly on course, by the appropriate application of stern power, or her movement may be accelerated by an ahead movement.

It is now proposed to discuss the cases of effecting a connection with a disabled vessel under the following four broadly stated cases:—

    (*a*) Vessel lying beam on to the wind.

    (*b*) Vessel lying obliquely to the wind.

    (*c*) Vessel lying head up to the wind.

    (*d*) Vessel lying stern up to the wind.

(*a*) **Vessel lying beam on to the wind.**

A relatively large proportion of ships lie beam on to the wind when stopped, but possibly the best examples are provided by the large passenger

K

ship type. This principally because of the huge, well balanced, windage offered by their vast superstructures and freeboards, but supported also by the fact that such ships are normally most meticulously trimmed and, although such ships are invariably multi-propelled, the screws are usually well forward of the after extremity so that drag effect is not great in relation to the other quite massive factors.

The larger plan feature in Figure 39(a) represents a large passenger liner of about 700 ft. in length stopped in the open sea in wind strength somewhat in excess of Force 4.

The tug commences operations by proceeding to windward of the disabled ship to take up a position allowing her to run down with the wind dead astern to pass very close under the stem of the ship. She should proceed as slowly as steering will allow just keeping as much ahead of the sea as will prevent momentary broaching. As the tug's bow passes the ship's stem, the initial connecting medium should be passed . . . (The actual connecting processes will be dealt with fully in a later section) . . . When this has been done, the tug being held under the stem if necessary, by the use of astern movements in the event of a missed line or other similar hitch, the tug should run slowly ahead so that her stern is just clear to leeward of the casualty's bow to await the process of boarding the towing gear.

If the correct position is achieved, i.e. just ahead of and to leeward of the casualty's lee bow, this position can be maintained indefinitely and with comparative ease because the leeway made by the disabled vessel will permit of judicious ahead movements to promote steering whilst the tug's known handling characteristic will allow of astern movements to maintain status quo.

When the weather is more than averagely adverse, this final phase of connection can be completed quite within the shelter of the casualty's bow, because the acceleration of lee drift occasioned by the weather will permit of easier station keeping by the tug. Needless to say the tug's towing deck party will appreciate this consideration.

After the connection is completed the tug should not attempt immediately to shape a course but should run on down to leeward paying out her towing gear. When this has been done the tug should take the full weight of towing, stretch the gear, and take up the requisite course in easy stages.

Regarding the proximity of the tug to the casualty at the time of passing lines; generally speaking the closer the better as it accelerates and promotes the passing of the connecting media and so reduces the strain on the working parties of both ships. It all depends however upon the weather obtaining and the sea conditions. If the sea and swell are running at any appreciable angle to the wind, dead ships lying in the seaway tend to take up a corkscrew motion which causes bow and stern to swing, or yaw, considerably in arc with a corresponding apparent ahead or astern movement, this can be deceptively dangerous and should be allowed for. (Figure No. 40).

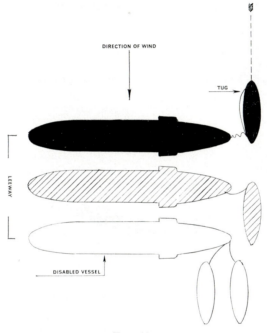

DIRECTION OF WIND

TUG

LEEWAY

DISABLED VESSEL

Fig. 40.

(b) Vessel lying obliquely to the wind.

It was observed in Chapter I of this Section that vessels lying obliquely to the wind do not move directly down-wind but make composite way depending upon the attitude taken up. A vessel constrained to lay head off the wind, due to her windage distribution, will make headway as well as leeway, and one laying head on to the wind will make some sternway. Taking the two cases separately, the following procedure is offered.

(b) Vessel lying head off the wind and making headway.

Having circled the casualty to estimate the rate and direction of lee drift, proceed to windward as described at (a) above, but instead of assuming a position dead to windward of the casualty, run past this position somewhat before assuming the down-wind approach attitude. Clearly no rule can be offered as to the amount that should be allowed seeing that this will vary, most considerably, according to the ship to be assisted and the weather at the time, but it is offered that, in this, it is wise to over estimate rather than otherwise seeing that the tug's astern working handling characteristic can be used in the event that the allowance has been too generous whilst an under estimate will obviously require a fresh approach.

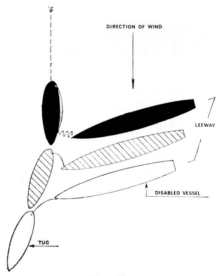

Fig. 41.

The tug may be stopped as her bow passes the casualty's stem, and as the latter bears down the lines may be passed, whereupon the tug should assume a course parallel to the line of drift whilst a full connection is made. As in the case (a), this whole procedure keeps the tug in safe relation to the disabled vessel with full freedom for manoeuvre in the event of untoward occurrence. The lee drift of the casualty also permits of judicious helm and engine movement in order to derive maximum benefit from the casualty's lee under inclement conditions. (Figure No. 41).

Vessel lying head towards the wind and making sternway.

This is perhaps the easier condition as the risk of collision is decidedly less, besides which this circumstance allows of an easier adjustment in the event that an inaccurate first appreciation of drift was made.

As described at (*b*) 1, a careful estimation of relative lee drifts should be made followed by the standard down-wind approach, lines being passed as the casualty's stem is passed with subsequent activities following the procedure outlined earlier.

In both of these cases due heed must always be paid, when making the approach, for the yaw which almost always obtains when vessels lie at broad angles to wind and weather, and which causes an apparent alternate ahead and astern motion which can be quite disconcerting if it has not been fully appreciated at an early stage in the operation. (Figure No. 42).

(*c*) Vessel lying head up into the wind.

Probably the most trying condition of all, and there is a temptation in such cases to try a parallel and bow to bow approach but, unless the approach is very close and very expert, and the performance of the tug's crew equally expert, this procedure is fore-doomed to failure because with the slightest hitch, the tug will not hold a head to wind attitude when stopped whilst the first touch of astern engines must induce an opposite attitude, which may well spell disaster if the tug's head should pay off in towards the disabled vessel.

The safest, if probably the more tedious, way to effect a connection under these circumstances is to commence operations by bringing a considerable length of towing wire up on to the tug's forecastle head from the towing deck, outside of everything, bridling the bight down hard right aft. Then make a down-wind approach, aiming to bring up on one or the other bow of the casualty, just off the true end-on position. When the tug's bow is just past the disabled vessel's stem, hold the tug with an adequacy of astern movement, and pass the lines. Then pass sufficient towing wire across for the casualty's forecastle head party to make fast temporarily, then allow

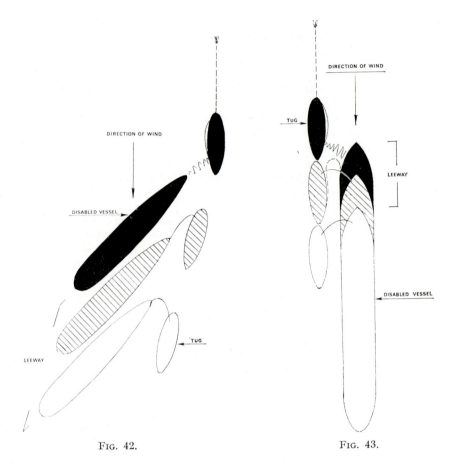

FIG. 42.   FIG. 43.

the tug to drive down-wind until the lead of the gear runs clear of the screw, then cast off the balance of gear from the tug's forecastle head. Thereafter hold the tug, using helm and engines as necessary, with her stern in close proximity to the disabled vessel's fore-end so as to not unduly stress the working parties of both ships, until a total connection is made. As before run down-wind with the tug whilst fleeting the gear and assume course by easy stages. (Figure No. 43).

A true head to wind aspect for a casualty is rare and more usually vessels having this appropriate windage characteristic lie a few degrees off the wind on one side or other. When this is the case, the better approach is usually towards the side which has the slight lee because the vessel's attitude will provide a little more room for manoeuvre whilst fleeting the gear.

(d) Vessel lying stern up into the wind.

This represents, relatively the easiest operation of all for the tug seeing that both vessels concerned have the same drift attitude. Initial estimations of drift should be followed by the standard down-wind approach, except that, in this case, there will be time and enough to spare and because both

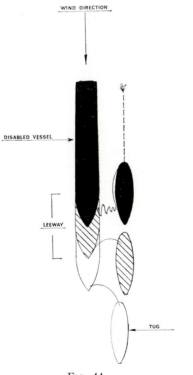

WIND DIRECTION

DISABLED VESSEL

LEEWAY

TUG

Fig. 44.

vessels are facing in the same direction and disparity in lee drift will be slight. The whole operation of connecting will be in full view of all concerned so that co-operation should be excellent. Fleeting of the towing medium should present no difficulty under such circumstances so that the business should proceed with celerity. (Figure No. 44).

(In the Figures Nos. 40 to 44, the shapes coloured black represent the phase of making the first connection. The phase of passing messengers, etc., is indicated by the shapes filled in with hatching and the wholly white shapes indicate the relative positions of the vessels whilst the full towing connection is made).

CHAPTER III.

The Practical Seamanship Involved when Effecting a Towage Connection between Ocean-Going Tugs and Disabled Vessels in the Open Sea.

1. Passing the Towing Rope.
2. Securing the Towing Gear Aboard the Disabled Vessel.

In the consideration of 1 above, it is offered that there are five methods of passing a line to a disabled ship, these are:—

        (*a*) The Hand-thrown heaving line.
        (*b*) The Line throwing Rocket.
        (*c*) The Line throwing Gun.
        (*d*) A towed line.
        (*e*) By Helicopter.

(*a*) The Hand-Thrown Heaving Line.

It is the practice in sea-going tugs to make up heaving lines for exclusive professional use. These lines are made up somewhat longer than those in common marine use being of from 40 to 45 fathoms in length, this in view of maintaining contact under marginal conditions. It is also the practice to finish up the ends of the lines by working an eye of about six inches length in to the inboard end in order to facilitate a marrying with a quickly formed sheet bend, and by ballasting the thrown end. Some tugs use a monkey's fist knot made up about a large hexagonal nut for this purpose, others utilise a small canvas sand bag for the purpose; eight ounces of sand make up into a handy flighting weight approximately 4 in. × 2 in. diameter, which when provided with a brass eyelet secures conveniently to most heaving line stuffs.

Before the advent of man-made fibres, the favourite flighting ropes were 1¼ in. to 1½ in. cotton or tarred hemp. In search of flexibility these ropes were usually thoroughfooted very vigorously so that it is probable that their actual strength in use was something appreciably less than was tabulated and rather less than was desirable in view of the function. Now however Terylene is the favourite providing all the flexibility required for true flighting together with a full adequacy of strength up to the practical requirements.

When making a first connection, the heaving line hand should be stationed on the forecastle head just forward of the bridge and he should be provided with *two* lines, one in case of accident. A third line should be run outside of everything from right aft up to the forecastle head guard railing, stops being used as and where necessary, this line being provided with a whipped end ready for marrying into the eye-splice in the thrown line.

As the tug comes down-wind to the disabled vessel, but whilst she is still a little to windward, the line should be thrown so as to land the flighting weight well on deck. The moment the line is *safely in hand* on board of the disabled vessel, the heaving line hand marries the thrown line to the one leading aft, throws it clear and indicates his action by a shout.

On the towing deck a coil of messenger line has been broken out and a soft eye has been worked into one end, whilst the coil has been flaked out on the disengaged side. As soon as the shout is heard; the line is bent into the messenger eye with a double sheet bend and the messenger is hove aboard the casualty. When sufficient messenger line is hove aboard, the messenger is cut and bent on to the end of the towing rope.

It will be observed that this drill does not employ a messenger cut to a specific length, but that a length is cut to suit each operation, this provides a double advantage. Firstly, under really good conditions permitting a really close approach a sixty fathom messenger might be so long as to provide an embarrassment whilst, under bad conditions, the same length might necessitate a marrying which, if it failed, might occasion an entirely fresh approach.

The common method of securing the messenger to the eye-splice in the towing wire is to pass the end *through* the eye and make a bowline, but when the eye is hove up to the fairlead of the casualty the greatest difficulty is experienced in boarding the eye because it will bind at the fairlead and although ratings reach outboard only one man can grasp the eye with one hand and he cannot alone lift it inboard. Therefore secure the messenger to the splice of the eye of the towing wire by means of a rolling hitch, then when it is hove up to the fairlead, the eye will stand up clear and can be conveniently brought inboard by reeving a rope's end through it which can be laid on to by several hands, or which may be brought to the windlass.

The actual throwing of heaving lines is rarely fully described so that no apology is offered for doing so here.

The line should first of all be thoroughly wetted and stretched, then about twenty fathoms of it should be made up into a conveniently small right-handed coil, the remainder being flaked down upon the deck clear of the thrower's feet.

The coil should then be divided into two so that rather less than a half is held in the right hand with the remainder in the left. The thrower should then stand with his left side towards the direction of the throw. Then with the left hand under the right swing both arms to the right and then back across to the left with a good strong and free swing, letting loose the coil in the right hand at the pitch of the swing, and allowing the coil in the left hand to run out under the thumb. A good follow through is essential.

The ratings should practice this until proficiency is gained and it is well worth any Commanding Officer's while to stimulate interest in this aspect of seamanship by any means to hand. Most seamen can throw a line 75 ft. to 80 ft., but a trained man can beat this by up to 50 per cent, especially when he is provided with properly considered equipment.

Without wishing to labour this throwing of lines unduly, attention must be drawn to the necessity for not bothering the rating entrusted with the throw with a deal of unnecessary advice or orders; having given precise instructions on what is required of him, the rating must subsequently be allowed full initiative in the timing of his throw. Heaving line throwing is very much akin to certain competitive field sports and it must be appreciated that, besides the heavy weight of responsibility that is resting with the rating concerned, he is required to accumulate a deal of nervous energy for the job, a situation which is definitely not assisted by a barrage of largely inane advice or instruction.

(b) The Line Throwing Rocket.

Because all merchant ships of 500 gross tons and over, and all naval vessels, are obliged to carry line throwing apparatus to certain well established standards, it is not proposed to repeat the very ample descriptions, which appear in all of the standard seamanship books, of the better known marks of these appliances such as the Wessex or the Schermuly.

In connection with the down-wind approach which this work recommends it becomes necessary, in the consideration of line throwing rocket gear, to offer two observations concerning the use of the gear on board of ocean-going tugs.

The first concerns wind effect upon the tail of line depending from the flighted rocket.   Rockets fired across the wind tend to head up into the wind whilst those fired down-wind tend to fly high.   *This must be anticipated and allowed for otherwise all of the range advantage deriving from the use of rockets may well be lost in terms of time wasted in disentangling the light line from superstructure, deck fittings or rigging.*

The second observation is relevant to the occasion when the casualty is a tanker because such vessels always provide an added hazard when they are laden with crude petroleum or with those refined petroleum products which have low flash points.   This, it must be emphasised, applies not only to dangerous liquids which may be leaking out on to the sea surface, but also to dangers arising from the emission of inflammable or explosive gas/air mixtures from the tanker's gas escape lines.

In every case of providing assistance to a tanker, the prudent tug Commanding Officer will always ascertain the precise conditions obtaining before he releases a rocket.

(c) The Line Throwing Gun.

Both the hand thrown line and the rocket line offer certain operational limitations in ocean tug usage; the former, at best, suffers in terms of both range and accuracy of direction whilst the latter provides certain complications, not the least of which is the trepidation which it inspires at the receiving end.   Clearly the exigencies of Ocean Tug operations demand an appliance which will provide an adequacy of range together with a fair measure of directional accuracy and full reliability.   In the search for such qualities, certain small arms manufacturers have interested themselves in the development of line throwing guns; evolving mostly from experience gained in the First World War in the production of the rifle grenade in its various forms and applications.

Early endeavours in this direction began with the firing of brass line carrying bolts from the muzzles of ordinary service rifles using service blank charges as the propellant.   The bolts used consisted of brass rods, of a diameter somewhat less than the bore of the standard service rifle and of lengths approximating to nine inches, in most cases.   These bolts were provided with a formed eye at their outer ends to which a light line could be attached by a splice or bowline.   This device would throw a line a distance of 60 yards under reasonable conditions.   Accuracy was however non-existent and performance was understandably rather erratic. Detail improvement followed, over the years, in terms of modification to the bolt and propellant but significant advance was not apparent until the advent of the Coston Line Throwing Gun during the inter-war years. Whilst even this appliance did not provide any marked increase in range, it did offer much enhanced directional efficiency and, because it was a functional appliance and not a modification of equipment away from an original purpose, it was convenient to use and it was entirely reliable.

Full satisfaction in this aspect of line passing did not obtain however until the advent of the Greener Line Throwing Gun. This is a single barrelled gun which very conveniently breaks down into two parts for stowage and portability, and which incorporates a strong and reliable Martini type

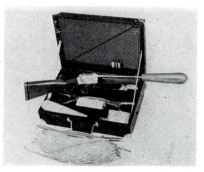

FIG. 45

action with an effective safety device. The barrel of the gun is ground truly parallel externally so as to accept a special line carrying projectile which fits sleeve-wise over the barrel. This projectile is of hollow form and streamlined shape for true flight; it is made of stainless steel and is cleaded with wood and finished to a conspicuous colour to facilitate recovery. The projectile is provided with a suitable lug at its rear end for the attachment of the line and is propelled by means of a specially made blank cartridge which breech loads into the gun.

The line, which may be obtained in a variety of sizes according to operational requirement, is made up into a non-fouling pattern by the use of a special winding frame. This frame associated with nylon line and a recoverable projectile makes for the ultimate in economy of operation. The gun offers the following performance:—

Effective Range in Yards	Size of Nylon Line in terms of Breaking Strain
55 yards	1,200 lbs.
70 yards	650 lbs.
90 yards	325 lbs.

Because the gun is very well designed and because the propellant charge has been carefully assessed, firing may be carried out from the shoulder position without discomfort and with full accuracy. The gun therefore provides an entirely practicable means of passing lines over the distances commonly associated with Ocean Tug operations. Because the projectile embodies no flame effect it is wholly safe in operations with tankers. (Figure No. 45).

(d) A Towed Line.

It is occasionally the case that a towage connection between an ocean-going tug and a disabled vessel cannot be achieved by means of a thrown or otherwise projected line. This sometimes occurs when adverse weather complicates an already difficult operation such as may present with a ship afire or, quite possibly, suffering some extensive structural damage resulting from collision or explosion. In such cases recourse must be made to a towed line.

A buoyant line of sufficient strength to messenger the towing medium from the tug to the casualty is an obvious requirement to such an operation. Only a few years ago this requirement provided the critical factor in this sort of activity seeing that the qualities of buoyancy desired were only provided by the Grass Fibre types of rope, types not particularly suitable to the crushing and torsional stresses set up whilst heaving substantial weights of steel wire rope inboard of ships. Today however no such complication arises because certain man-made fibres, such as polythene and polypropylene, are fully floatable and are hardly less strong, size for size, than nylon or terylene.

The actual process of floating a line calls for a float, at the tail of the line, of such bulk as will offer sufficient resistance to forward movement as to allow for positive control besides providing a plainly visible terminal marker. Some ocean-going tugs keep a specially made float for ready use in this capacity which usually takes the form of a small wooden barrel or steel drum of about five gallons capacity and provided with a steel wire strop and beckets for attachment to the line. Other tugs, considering that the use for such an appliance obtains but rarely, are quite content to improvise as and when the need arises.

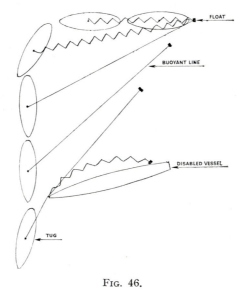

Fig. 46.

The drill for making a connection by this means consists of preceding the standard down-wind approach by running across the wind a cable or so to windward of the casualty. The float is then dropped overside in such a position as to allow sufficient scope of line according to the attitude of the disabled ship to the wind. When the correct position for a down-wind position is reached, the line should be turned up inboard of the tug to the towing hook or thereabouts. The line will then trend off somewhat abaft of the beam, and if the cross-wind run has been nicely judged, this trend will only have shifted aftwards by a point or two when the tug passes under the casualty's stem. At this juncture the tug should be stopped to allow the line to fall slack and allow the float to drop alongside of the disabled vessel where it may be grappled for as convenience allows along the ship's side.

Thereafter the normal sequence of boarding the gear ensues followed by streaming the towing gear and assuming the towage course. (Figure No. 46).

(*e*) The Helicopter in the Towage Sphere.

The well-read seaman will doubtless have read of the exercises which have been conducted by the navies of the world using helicopters in the role of tug, and will have made his own estimations of the practical values of such, There can be no doubt though of the contribution which rotating wing aircraft can provide in assisting orthodox towage operations by passing the gear. Helicopters are entirely capable of lifting considerable weights and of performing very demanding evolutions under trying circumstances whilst so laden, so that the transport of towing medium between a tug and a casualty up to a distance of 300 yards is entirely within their capacity.

CHAPTER IV.

Securing the Towing Medium to the Disabled Vessel.

It is most earnestly recommended that, whenever circumstances will allow, the tug's towing medium should be secured to the anchor cable of any casualty requiring towage assistance, this for four excellent reasons:—

1. A ship's anchor cable is the strongest flexible medium in the ship, having been designed to hold the free floating weight of the vessel under any normally conceivable circumstances of wind and weather.

2. A connection is rapid, simple and economic of material once the anchor has been removed.

3. The windlass whether powered, or in hand gear, provides an excellent means of effecting adjustments to total towing length to suit the changing conditions of towing.

4. Inboard chafe to the towing medium proper is wholly eradicated and is absorbed by windlass, cable and associated fittings specifically designed so to do.

It is well worth while going to endless trouble to achieve this means of connection even to the point of accepting a certain delay to the initial connection to ensure it.

The ideal method of utilising the full benefit of the facility is to disconnect one anchor from the end of its cable and to remove it wholly inboard. This provides the whole complex of windlass, cable, cable compressors and hawsepipe available for towing. Unfortunately this is not always possible when heavy weather makes the manipulation of a heavy anchor, by winch and derrick, an impracticable operation and it is entirely understandable that many a shipmaster will feel disinclined to sacrifice an anchor to the sea in order to derive advantage which may not appear as necessary to him as it will to the tug Commanding Officer.

An alternative in such cases is provided by securing one anchor in the pipe by the use of the cable compressor and such other auxiliary fittings as are available, thereafter disconnecting the cable at its first inboard shackle

L   149

and then leading the cable out from the windlass gypsy through the forward lead which gives the best obtainable line. This offers a reasonable substitute for the best but the line from the gypsy to the fairlead utilised must be most carefully considered and any sources of friction provided with adequate chafage gear. It may also be necessary to provide shoring and stiffening to the forecastle bulwarking in the vicinity of the lead.

The actual connection between the towing medium and the cable end must be made with a cable connecting type shackle with an oval sectioned bolt secured with a flush-fitting tapered pin held by a lead pellet, or its equivalent. Screw shackles should not be used because risk of stress distortion may well cause the bolt to seize with consequent ultimate embarrassment for all concerned. There is also the possibility that a screwed shackle pin might work loose.

When, for any reason, a cable end is not available then recourse must be made elsewhere and the first and obvious choice must be the forecastle head mooring bitts. The use of three sets of bitts is recommended as an effective connection in this case. The procedure usually followed is to take a right-handed full turn around the forward member of the pair of bitts in the truest run to the selected fairlead, following this with two full cross-over turns. The end should then be taken to the second pair of bitts turning it up there with three turns, similarly about a third pair of bitts.

This procedure is sufficiently friction provocative to allow each bollard to accept its fair share of the load but it may be found necessary to use a mallet, as the tug takes the towing weight, to ensure this. When the wire is fully settled the turns should be lashed at each pair of bitts with small chain and a prudent officer may well consider it desirable to lash the end eye of the towing gear to a convenient deck eye bolt.

The bitt equipment of most modern vessels is entirely adequate to the stresses of towage; when however the age, or general condition of a ship would appear to indicate the need for caution then some shoring of the bitts against the direction of strain might be undertaken.

In the smaller classes of vessel, whether merchant or men o'war, and where no special towage arrangements have been made, it is not uncommon to find that the bitts are insufficient to accommodate the size of gear in use, and it occasionally becomes a matter of some difficulty to find a suitable location for making fast the gear. When such conditions do arise it is recommended that a two, or four, part strop might be made from the ship's own cable sufficient to pass around the windlass, a gun mounting or even a deckhouse. It is, in this connection however, unwise to assume that a power-driven vessel's masts are as effective in the towage role as those of sailing vessels.

In all cases of towing by this method the crew of the disabled vessel, or the boarding party, must use full initiative in the arrangement of protective anti-chafe material as and where it becomes necessary both inboard and outboard at the outer parts of the hawse-pipe and the stem; imaginative construction here allied to conscientious subsequent supervision can greatly ameliorate an otherwise wholly unsatisfactory condition. Finally, an inboard connection of this type should only be regarded as suitable to a relatively short term towing operation.

The chief objections to the method are:—

1. The towing wire is subjected to the very severest crushing and torsional stresses at the fairlead and at the bitts. These stresses will almost certainly be more damaging than the straight stress of towage itself.

2. So much wire has to be taken inboard to allow for securing properly and to allow for 'inching out' to alleviate local damage, that the whole phase of passing gear becomes so protracted as to place the tug in hazard.

It is commonly the practice among the more prudent tug officers to shackle a pennant of suitable length to absorb all of the punishment imposed by an inboard connection to the end of the towing wire to be used. It is also a common practice to use a pennant somewhat larger in circumference than the rest of the equipment in view of the physical punishment suffered.

For a great number of reasons, all of which are entirely obvious, disabled vessels should be towed bow first. There are however cases where this is either impracticable or unsafe. After collision damage for instance, it is not uncommon to find a ship trimmed well by the head with no means of altering the condition. Such a ship simply would not follow bow-first and must therefore be towed stern first. The same applies to a vessel which has severe damage forward and is open to the sea.

When a stern-first tow is intended, it must be anticipated by all concerned that the whole operation will be extremely difficult. Most ships, or other floating objects, in tow have a tendency towards yawing to a greater or lesser extent, but vessels towed stern first invariably yaw badly. If towage is contemplated in terms of any period of consequence . . . (and in this context it would be prudent to estimate this in terms of speeds in the order of two to three knots) . . . the ship must be towed on a bridle.

A stern towing bridle must not be made up on a bight but must consist of a pair of carefully matched bridle legs each secured to three pairs of bitts on each side of the ship and most carefully connected together, and to the towing medium, so that the end links and connecting shackles bear fairly and properly upon bolts and the crowns of links and not unfairly.

The legs of the bridle should properly be made up of lengths of anchor chain, but it is appreciated that, for a variety of reasons, it may not be possible to bring cable aft in the quantities required and in such cases the use of insurance wires is permissible. The same anti-chafe precautions as were suggested earlier are equally necessary when towing from aft, and the provision of a tripping wire of adequate proportions will prove to be a blessing when setting out on the tow and again when the operation is in its concluding phases.

Whenever a disabled ship's anchor cable is utilised for towing care must be taken to preserve one anchor, at least, for use in an emergency or for use in the event of incidental delay at the close of operations. It is also seaman-like prudence to allow that there will be one large wire available for use with the displaced anchor, or the spare, in case of need.

CHAPTER V.

Rendering Towage Assistance to an Anchored Vessel.

Tugs are, from time to time, summoned to the assistance of vessels lying to a single anchor or to a moor, disabling conditions having occurred whilst the vessel was in a suitable depth of water and in congenial circumstances for anchoring, or alternatively when anchoring may have been something of an emergency obligation following an engine failure, or kindred complication, whilst in close proximity to the shore or shoal water.

In calm weather this condition presents no difficulty seeing that the tug can proceed directly alongside, preferably of course on the side on which the tow-rope is to be secured, to make fast temporarily whilst the lines are passed and a connection effected. The casualty can, at the same time, shorten cable and make all ready for getting under way so that the tug can cast off and run ahead, fleeting the towing gear, as the anchor is weighed. There will then be the varying degrees of deterioration in weather which will make this procedure progressively difficult until the position is reached when any contact between tug and casualty becomes increasingly hazardous to both. In such case the tug will find it difficult, or quite impossible, to hold station near the casualty for the period required to pass lines and in this case, sea room permitting, the casualty should weigh anchor to allow the tug to make a suitable approach according to the relative lee drift characteristics of the two vessels.

Clearly, circumstances must arise when a disabled vessel cannot weigh anchor to facilitate a connection because of a close proximity to a lee shore and when, in all seamanlike prudence, anchors cannot be weighed until the tug has the weight. In such case, and if the casualty's anchors are holding and there is no immediate danger, equal prudence would indicate the desirability of removing the tug to a safe anchorage where the advent of improved weather may be awaited whilst keeping a vigilant watch upon the casualty.

The reason for this apparent trepidation is that vessels at anchor in open water, under adverse weather conditions, are in a state of constant movement, a movement which, whilst it is constrained to a certain arc and area by the scope of cable and wind and tide, conforms to no predictable pattern; this behaviour deriving from the inconstancy of wind force and direction.   This erratic movement, added to the more normal difficulties resulting from foul weather, renders good station keeping quite impossible, whilst there is a grave risk of fouling the tug, or her propeller, or the towing medium on either the casualty's anchors or cables whilst passing lines or fleeting gear.   Towage and Salvage History notes repeated instances of collision and structural damage, often accompanied by serious injuries to involved personnel, resulting from the mishandling of this seemingly quite unspectacular operation.

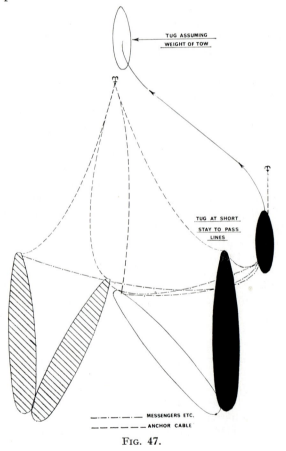

Fig. 47.

When imminent risk to life and limb demands that attempts be made to assist a distressed vessel under these circumstances, the tug should approach with all caution to endeavour an estimate of the lay of the ship's cable to her anchor or anchors. She should then move in close to the casualty whilst she is on a suitable sheer and pass a line, by any of the means described earlier, thereafter moving a little outside of the estimated arc of sheer to anchor using the shortest scope of chain practicable to the conditions.

The gear should then be passed from this position. There will, quite inevitably, be delays and difficulties but the fact that the operation will take place in shallow water precludes the risk of the gear taking charge. Most tugs are moreover sufficiently manoeuvrable to permit of a fair degree of controlled movement whilst riding to a short scope of chain and this is probably the only means by which head up to the wind sternway may be achieved with an ocean-going tug such as will be obligatory if the casualty should be dragging her anchors.

When a full connection is effected the tug should weigh anchor and run out in a wide arc to assume an ahead towing station in order to clear anchors or cables. The gear should be fleeted against the winch brake if a winch is fitted, or through a stopper if not, in order to keep the gear as close to the surface as possible until the ahead station is reached. The tug must then keep positive manoeuvrability thereafter whilst the casualty's anchors are weighed. This may not be altogether convenient to the disabled vessel's forecastle head party, but all hazardous rescue operations entail a certain degree of ancillary difficulty and hardship and always, once the tug has the weight of the tow, the casualty is in a position to sacrifice her anchors and cables to the sea in the greater interest. (Figure No. 47).

In the case of a vessel in such hazard lying to two anchors it is unlikely that the Master will be inclined to weigh one of them to convenience the tug's manoeuvring; in any case it is possible that this action may not produce unqualified benefit, seeing that sheering will probably increase.

CHAPTER VI.

The Operation of Shortening in the Tow.

On approaching its destination it becomes necessary to reduce the over-all length of a towage unit in order to promote manoeuvrability to facilitate the business of bringing the tow to an anchor, handing over to local tugs for berthing, or accepting extra towage assistance whilst traversing a greater or lesser distance of esturial or river steaming, or other port steaming, prior to an ultimate hand-over.

When an ocean-going tug is equipped with an automatic towing winch the operation of 'shortening in' is simplicity itself. The procedure consists of advancing the Winch Automatic Control to the top setting and then progressively reducing propeller revolutions until the Automatic Control heaves in steadily, the controller should then be moved to the Off position and the Hand Controller engaged, the automatic winch brake holding the weight of towing whilst the transition is effected. The Hand Controller may then be used to control heaving in, the speed of towing being adjusted to the winch heave-in rate to obtain the desired rate of shortening in. When the towing medium is reduced to the desired length, the winch can be restored to the automatic function and speed may be resumed as required.

When no winch is fitted, and towing has been conducted from a hook or bollard, using a composite towing medium which includes a length of heavy fibre spring, then the operation becomes a vastly different proposition and requires team-work and good organisation if the whole business is not to be unduly protracted.

Given the sea-room and a clear sea bed, i.e. free from such submerged obstructions and complications as are presented by wrecks, rocks, cables and moorings and the like, the towage unit may be brought to a standstill in easy stages and the towing medium allowed to fall to the sea bed. When such conditions are not present however, or when strong current or tide or other complications presents itself, the evolution must be completed without losing way. These two cases must be dealt with separately, but both require the following items of gear:—

(a) A heavy messenger, tailored to the relative positions of the capstan or winch, molgoggers and fairleads, of about 4½ in. manila (36 mm.

dia.) or $3\frac{1}{2}$ in. Terylene (28 mm. dia.) fitted with a thimbled eye-splice at one end into which is worked a chain stopper $\frac{3}{8}$ in. dia. (16 mm.) provided with a substantial hook.

(b) A stopper of $4\frac{1}{2}$ in. Manila (36 mm. dia.) or $3\frac{1}{4}$ in. Nylon (26 mm. dia.) (10 mm. to 12 mm.).

(c) A $\frac{3}{8}$ in. to $\frac{1}{2}$ in. chain stopper at least 9 ft. (3 metres) in length fitted with a rope tail.

A. Shortening in a Composite Towing Medium With the Unit all Stopped.

Ease down towing speed in plenty of time so that the tow will not over-run the tug. When all is stopped lift the inboard part of the towing medium into the appropriate molgogger depending upon lee drift of tug and tow. Take the messenger aft, reeve it through the molgogger and make the snotter fast to the towing medium as far outboard as can be reached. Take the messenger to a forward lead and back to the capstan and commence heaving in. When a full fleet has been brought home, stopper off, using rope or chain stopper according to whether the inboard section of medium is wire or manila and according to length, and repeat.

It will be seen that the operation as described requires five ratings:

(a) A stopper hand.

(b) A hand to pass and make fast the messenger.

(c) A hand to control the capstan.

(d) Two hands to haul off the capstan and fake down.

If the capstan fitted will not accept at least three full turns of the fibre spring then the whole process of boarding the spring must be effected by repetition of the fleeting process until the wire pennant is reached when this may be taken directly to the capstan.

If the capstan is of a size as to accommodate a sufficiency of the fibre spring then the whole process is greatly accelerated, but a fair and proper lead from the molgogger to the capstan must be arranged. Ocean-tug capstans must not exceed the height of the bulwarks for the most obvious reasons whilst the molgogger must, for equally obvious reasons, be higher.

It follows therefore that a straight lead from the molgogger to the capstan must cause fouling if the spring is taken to the capstan in the normal fashion of hauling off part uppermost. It follows therefore that the turns must be placed upon the capstan in reverse order so that the feed is at the top and the haul off is done from the bottom. Enough spring should be fleeted inboard, as before described, to make three or four full capstan turns, then it should be stoppered off. To rig the turns, coil down the required number of turns on the deck adjacent to the capstan in the form of a large loose coil starting the coil from the outboard part of the fleet. Then lift the top coil off the pile and place it on the capstan, turning it over in the process. Repeat with the other two or three. When the last is in place it will be seen to lead off from the top of the capstan to the molgogger whilst the hauling part will lead off forward from the bottom. (Figure No. 48).

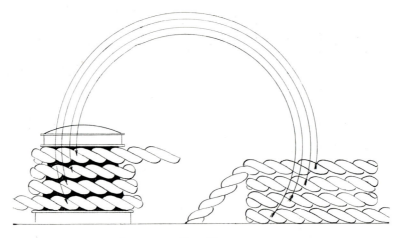

Fig. 48.

When this has been done, take the weight upon the capstan, come up on the stopper and heave away. When the tow is sufficiently close up, stopper off and turn the gear up on the towing bollard if fitted, if not drop the bight over the towing hook and either take it to the bitts or secure it to its own part with a seizing.

B. Shortening in a Composite Towing Medium Whilst Retaining Way on the Tow.

This can only be carried out given two conditions:—

- (*a*) Calm sea with no more than slight swell.
- (*b*) Sea-room to permit the tug to hold a steady course, or a near approximation thereto, for the duration of the operation.

These two conditions are fairly self-explanatory, the stopper could not hold the weight of towing if the effects of sea and swell were superimposed, whilst if course were adjusted to any extent, the molgogger would be obliged to act as a fulcrum to the towing medium, a stress which it is not designed to support. Given these two conditions, the job proceeds as before described except that the towing deck team is now augmented by two, viz. the Commanding Officer and the Quartermaster.

The Quartermaster must steer a meticulously steady course so that the towing medium shall traverse the molgogger rollers without weight or binding. The Commanding Officer must regulate the engines to suit the rhythm of the operation thus:—When the stopper is in use he must stop his engines to reduce the load until the capstan takes the weight, when he can build up revolutions until the end of the fleet is reached and the stopper is set up again. This cycle must be repeated throughout the fleeting-in process in order to keep the unit manoeuvrable and under command, also to ensure that no sudden or undue weight is thrown on the stopper to cause a failure.

In the case that the manilla or other spring can be taken directly to the capstan after the initial fleeting, then the Commanding Officer can proceed at a steady speed commensurate with the power of the capstan.

CHAPTER VII.

The Ocean-Going Tug in the Close-Quarters Role.

It is an unfortunate but entirely unescapable condition of specialisation that the products of this principle have only minimal functional value outside of the particular sphere for which they were developed: this statement is quite axiomatic in its application to every aspect of marine towage. The original basic tug type, since its appearance upon the nautical scene more than a century ago, has been separately developed to serve the particular requirements of specialist operators so that although each serves its allotted function faithfully, its performance in other aspects of the calling declines from the merely indifferent, in the case of functions akin to its own, down to a condition of virtual ineffectuality in those more distantly removed. In the case of the ocean-going tug type this functional distinction must be wholly appreciated because this class of vessel is, only too often, called upon to perform duties quite alien to its designed purpose.

Generally speaking the ocean-going tug is too large in every dimension, of too great a displacement tonnage and too slow in manoeuvre, to be employed in close quarters work and should only be considered in this connection in emergencies. *Having made this categorical statement, it is now necessary to consider the utilisation of these tugs in this particular role seeing that, as every Tug Commanding Officer knows only too well, the tug will, sooner or later, be required to work at close-quarters,* usually by persons with authority, small responsibility and less practical knowledge.

In particular reference to the broad classification of ocean-going tug types offered in Chapter II. Section 3, it is submitted that whilst vessels of Class 4 might well return a reasonable performance in close quarters towage assistance, those in the other Classes will be progressively less flexible. This is not to imply however that size is the only criterion in the matter seeing that this may well be qualified by other features of design or equipment. The content of this chapter applies none the less more to Classes 1, 2 and 3 than to 4.

Because of the foregoing, prudence would indicate a restriction of the ocean-going tug type in close-quarters work to the following conditions:—

1. Employment in the leading tug position *only* whilst towing by a tow rope from the towing hook or towing bollard.

2. Towing whilst lashed alongside.

3. A severely limited application in the steering function.

1. Employment in the Leading Tug Position.

This is the service commonly demanded of tugs at the termination of towage operations and it is one that can only be accepted subject to the following conditions:—

(a) The close quarters operation should not be unduly complicated, neither should it obtain under conditions of traffic where the tug herself will be placed in hazard in adopting any of the attitudes necessary for the manipulation of the tow.

(b) The unit of disabled vessel and attendant tugs should always proceed at slow speed so that the lead tug may keep positive weight upon the towing gear in order to maintain full manoeuvring initiative.

(c) Every major manoeuvre, or alteration of course, should be preceded by reducing the headway of the unit through the water to a minimum so as to avoid girding the lead tug.

(d) In view of the size and power of an ocean-going tug in the lead role and the momentum entailed in providing headway and steerage, the tug must be provided with some form of shock absorbing medium at the point of tow.

(e) A composite docking spring should be used, part steel wire and part terylene, of a strength at least 50 per cent in excess of the tug's bollard pull. The whole spring to be of such length as to provide a tug's length of clearance between the tug's stern and the assisted vessel's stem. The terylene spring to be of such length as to bring the connection to the pennant inboard of the tug.

(f) The docking spring should be bridled by a wire, manila or nylon hauling line, on the bight, to a bow shackle at the connection between the terylene and wire parts of the towing spring from the capstan, via a gog-eye with a hand in attendance throughout the operation.

(g) The slipping arrangements at the towing hook must be in first class order, and a second hand must be stationed there with an helm commander ready to slip.

2. The Ocean-Going Tug Towing Alongside.

This is the most satisfactory close quarters employment for the larger tugs because in this position the advantage of their great power is fully accessible without the disadvantage of their bulk in manoeuvre. Properly used and effectively secured alongside, large tugs are fully competent to single-handedly manoeuvre vessels of up to four times their own displacement tonnage.

The ideal position for alongside towage is fairly well aftwards on the ship to be assisted so that the tug's stern overlaps the other vessel's stern somewhat. This allows of a free and full run of water to the tug's screw and to the rudders of both ships. It also permits of a free run of propeller wash from the tug without reaction from the ship being towed. This position is readily achieved in most loaded ships of modern cruiser stern design, but becomes more difficult of acquisition in fine lined ships in the ballasted condition. In any case, the more afterly the position secured, the more effective the tug.

To make the best use of her power the tug must be properly secured alongside with a good head-rope to transmit stern-power, a good stern-line to apply ahead-power and a sufficiency of breasting to hold the tug close alongside. The headline is best taken away from the off-side bow and given as much drift as possible. The after line may conveniently be taken to a position well aft on the tow from the tug's towing hook. The breast-ropes should be so set up as to induce a slightly head-in attitude on the part of the tug so as to reduce the volume of water passing in between tug and tow and which tends to force them apart. When towing alongside it is important to have all of the gear set up taut so as to maintain full control and also to prevent surging which can very rapidly induce a break-down in the shorter length of breasting. To ensure this make a slow ahead movement after completing the initial make fast, this will stretch the after gear and cause the forward lines to fall slack, holding the tug alongside of the tow by helm action and with engines turning over ahead, take up all slack and set up again.

In subsequent manoeuvring it must be remembered that ahead movements by the tug will cause the tow's head to pay off away from the tug so that the rudders of both vessels should be adjusted accordingly. Conversely an astern movement will bring the tow's head over towards the tug's side. (Figure No. 49).

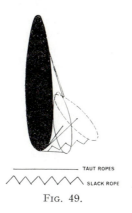

TAUT ROPES

SLACK ROPE

Fig. 49.

If the towage manoeuvre includes an ultimate berthing, then the alongside towing position allows of a ready transition from parallel towing to the 'butting' aspect by the use of the appropriate helm action in conjunction with a casting loose of the after breast-rope. It is equally easy to resume the parallel position. It is the practice in some tugs to take the after breast rope to the capstan and secure it there. Whilst most tug's capstans are more than adequate to the strain imposed this is not good practice and no delay of consequence is occasioned by taking the breast-ropes to the bitts.

The use of one side of the tow or the other for the purposes of alongside towing will usually be decided by the condition of the tow's berth upon arrival or at departure. Some tugs are especially fitted for alongside working and it would naturally be prudent to use the specialist equipment if available. When a choice obtains and when the towage operation includes river or channel steaming it is a great convenience to secure to the tow's port side so as to achieve the best view of oncoming traffic.

3. The Ocean-Going Tug in the Close Quarters Steering Capacity.

Steering a disabled vessel by the use of another vessel's engines and rudder is never easy, it is rarely truly effective, but it is an operation which must be endeavoured from time to time, usually under less than favourable circumstances.

The best way of steering a disabled ship is to lash a tug to each quarter to maintain command whilst a sufficiency of other tugs provides the forward movement. The next best way is to employ a tug, facing opposite to the direction of advance, applying turning power through the medium of a very well bridled tow-rope, but this method is clearly not applicable to anything lengthier than a dockyard or a port shift and is, in any case, inappropriate to the larger classes of tug.

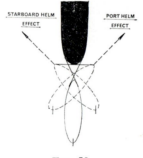

Fig. 50.

When only one ocean-going tug is available for steering the best attitude is with the tug heading in the same direction as the tow and secured thereto by the shortest drift of elastic towing medium which circumstances will allow.

When the lead from the tug's bow to the assisted vessel's stern trends downwards excellent control is provided by the use of twin ropes, one leading from the tug's starboard bow to the tow's port quarter and the other from the tug's port bow to the tow's starboard quarter. Control is then exerted by sheering the tug out to the quarter opposite to the direction in which it is required to steer the tow. The crossed lines allow the tug to exert the requisite power with relatively small changes in relative position whilst the application of power to the inside quarter of the tow, when inducing swing, provides superior mechanical advantage to the same power applied to the centre line aft. (Figure No. 50).

This method cannot be utilised, of course, if the trend of the lead is upwards from the tug seeing that, in such case, the stresses involved must necessarily be supported about the tug's forecastle head bulwark plating. In this circumstance a single steering line must be used in conjunction with the bow-port and the forward towing post, the increased translatory motion required to produce equal steering effect being accepted.

M

CHAPTER VIII.

Yaw whilst Towing.

Yaw in vessels, and other floating objects, being towed is the rule rather than the exception so that a chapter devoted to this pernicious concomitant to most towage operations cannot come amiss. Yaw is a worry to any Tug Commanding Officer, not only because of the considerable reduction to the unit's speed which is the first of its products, but because of the extraordinary stresses which such behaviour imposes upon part of the towing medium. The loading upon that part of the towing medium which makes first contact with the vessel or object towed is quite enormous when the tow brings up to the gear at the end of each run, besides which the additional chafe, imposed upon the towing medium at both ends as a result of the continuous and substantial lateral movements of the tow, provides a separate problem of no mean proportions.

Yaw may be caused by a variety of reasons of which the following are the most common. These may be experienced separately or in any combination depending upon the nature of the disabling factor in a casualty, or in peculiarities of underwater form or trim, or superstructure rig, in any floating object in tow.

- *A.* Low Speed in Tow.
- *B.* Unsatisfactory Trim.
- *C.* List.
- *D.* Disposition of Superstructure *v* Wind.
- *E.* Immersed or Part-Immersed Hull projections resulting from Damage.
- *F.* Too High a Towing Speed.

In each case of a badly yawing tow the Commanding Officer must endeavour to identify the cause, or combination of causes, which produce the ill-effect and thereafter take steps to remove or diminish them. It must be stated at once however that in the majority of purely rescue or salvage tows, particularly if the vessel is a hulk, there is very little to be done as invariably such vessels include a loss of power, both main and auxiliary, as a prime part of their condition.

A. Low Speed in Tow.

It is apparent that unless a vessel can be towed at a speed at least equal to her normal steering speed . . . (Viz. the minimum speed at which any vessel can proceed so that she will answer her helm) . . . she will yaw and nothing will prevent this except, of course, the expedient of increasing her speed up to the practical minimum. This may be achieved upon occasion by the disconnection, forward of the thrust block of course, of the tail end shaft from the main engine to allow the propeller to revolve freely. The effect of this proceeding is considerable as is shown in the paper 'Permanent Moorings,' by Messrs. Thorpe and Farrel, where the propeller resistance to forward movement is:—

$$\text{Propeller Resistance in Lbs.} = \text{Propeller Diameter in feet}^2 \times \text{Speed in knots}^2 \times 1 \cdot 43$$

although in his paper 'Some Observations on Towing Tensions,' Captain J. Logan did suggest that the co-efficient $1 \cdot 43$ was on the high side for most merchant ships.

In the event that this action does not produce a sufficient increase in speed to ameliorate or remove yaw then the only other recourse open is to augment towing power by the addition of more tugs.

B. Unsatisfactory Trim.

The Department of Naval Construction has offered certain recommendations on yaw resulting from unsatisfactory trim, and in view of massive evidence provided by the records of, quite literally, thousands of tows, these recommendations must be accepted as being most authorative and competent therefore to clear up a deal of uncertainty on those points.

Positive trim by the stern is regarded as being entirely essential but trim by the stern should in no case exceed the forward draft of the vessel. The following table offers examples of trim which are experience proven:—

Displacement Tonnage of Vessel	Trim by the Stern in Feet
Up to 1,000	1 foot
1,000 to 7,000	2—3 feet
7,000 to 15,000	3—6 feet
Over 15,000	4—8 feet

In small fine lined vessels it is necessary to make an allowance for the weight and effect of the towing medium.

Objects in tow of approximate rectangular plan form, with flat ends and bottom, also follow more steadily in tow when provided with a slight trim by the stern. Dutch Tug Masters, with great experience in dry dock towage, usually commence operations with an allowance of afterly trim which approximates to 3 in. per 100 ft. of length, thereafter adjusting from this trim if the tow does not follow well.

Trim can, of course, be altered by flooding or pumping out appropriate water-tight compartments, but this is something to approach with some caution in a damaged vessel, and unless the casualty has auxiliary power enabling the necessary corrective action to be taken, flooding should only be carried out in the last recourse. When flooding remains the only and necessary corrective, preference should be given to small compartments well below the water line so that they may be pressed full, this will avoid hazard due to large masses of slack water and the attendant adverse effect upon the casualty's metacentric height due to free surface.

C. List.

List causes yawing because of the imbalance of immersed and exposed areas. In calm weather listed vessels tend to run off towards the lowest side, bringing up to the tow-rope with reluctance and thereafter sheering back until the weight of towing brings them in line once more when the unbalanced forces generate another sheer *ad infinitum*.

This condition is greatly effected by the wind. Stern winds and head winds tend to aggravate yaw due to list, a beam wind on to the low side will reduce the sheer but wind upon the high side will increase it.

Once again the palliative action is to reduce or remove the cause by some alteration to the vessel's lading or ballastage, but again this will have to be approached with great caution.

D. Yaw Due to Wind on the Superstructure.

The most troublesome tows, in this case at least, are those with the preponderance of their superstructure areas forward of amidships, from any point of the compass wind effect is bad in ships so constructed, but with strong winds blowing at a broad angle to the course such vessels can become quite unmanageable and the only action open to the tug is to adjust course and speed to reduce stress upon the gear.

Vessels with their superstructure grouped about the after end are kinder to the towing medium in that they usually run off a little to leeward until the wind pressure and towing medium stresses balance, and will hold such a position so long as the balance maintains in terms of wind direction and towing stress. Fair winds are occasionally troublesome but not extremely so.

E. Yaw Due to the Effects of Immersed or Part Immersed Hull Projections Resulting from Damage.

Correction of this type of cause is usually quite outside of the scope of tug crews at sea. Very calm weather may permit of some burning or controlled cutting with explosives but not to any appreciable extent. Usually the only corrective action is to try to rig sufficient windage to compensate for the imbalance caused by the projecting damage.

F. Yaw Due to a High Towing Speed.

Speed in tow is frequently a critical factor with yaw. When the prime cause of yaw results from trim or list, and it cannot be corrected, it is sometimes found that an alteration to the speed will help matters. In the case of trim yaw a speed reduction will invariably help, but this is not the case when list is the cause because practical experience has shown that an increase of speed will sometimes hold a towed vessel out on a maintained sheer for days at a time.

Generally speaking a constant sheer to one side or the other is less bothersome to the tug. This condition does, of course, reduce speed but if a steady sheer can be promoted, particularly if it can be induced so as to give a non-chafe lead for the towing medium, it is less stress provocative than the uncontrolled yawing. If the casualty's rudder can be hand operated, fifteen to twenty degrees of helm can often help in this way, but more than twenty degrees should not be used as it may induce yaw through the reduction in speed occasioned. In any case sheer induced in this way will only maintain as long as the conditions of sea, wind and speed through the water remain

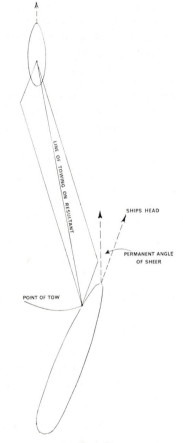

FIG. 51.

constant; any variation in either of these will necessitate a fresh adjustment so that the utilisation of this measure predicates the manning of the towed vessel.

When the rudder is immovable the same effect of constant sheer can be obtained by making the point of tow in the casualty at some position on the shoulder.   Once under way the common parallelogram of forces system exerts which produces a tendency to sheer the vessel away from the point of tow, the angle of sheer being directly proportional to the distance of the point selected from the stem head.   (Figure No. 51).

The limitations of the method need no advertisement however and it can only be applied in the case of relatively short tows under good conditions of weather or in sheltered waters.

When the Beam to Length proportion of a floating object towed is high a towing bridle will impose a large measure of control upon any tendency towards yawing but this ratio, in most merchant ships, and all warships, is too small for any positive effect.

Whilst experience of ocean towage has shown that the rigging of sails, either forward or aft, has produced beneficial effect in counteracting yaw from different causes, there seems to be little evidence that the towing of drogues can, in any practical manner, effectively assist matters.   In the case of either application the presence of a crew on board of the casualty is clearly necessary, a condition which cannot always be satisfied.   In any event a drogue of sufficient size to produce any effect upon even the most modestly sized merchant ship would require some auxiliary assistance.

In conclusion, it is submitted that when a yawing tow is very troublesome, and when the economics of the operation permit, the only measure likely to provide an enduring solution is the employment of a second tug to steer from aft.   In this attention is drawn to the use of the cautionary expression: 'likely to,' seeing that, in almost every case except that of yaw caused by trim, the best place for a second tug will probably be up beside the first one.

CHAPTER IX.

The Beneficial Effects of Oil whilst Effecting a Towage Connection at Sea and whilst Towing.

Before expatiating upon the beneficial effect of oil upon rough sea, one must make the reservation that, whilst this initial effect is not denied, the combination of oil with sea water can produce difficulty in the subsequent man-handling of lines and hawsers, and that the mixture upon the deck of an ocean tug, in some movement in a seaway, can be quite dangerous and exceedingly uncomfortable. The initiative to use oil when effecting a towage connection at sea is therefore one to implement only after a very great deal of consideration and, in any case, an initiative resting wholly with the tug.

If it is considered advantageous to spread oil then, because the disabled vessel is drifting to leeward and because the tug's approach will almost invariably be from the windward, then the oil must be spread from the vessel requiring assistance. The oil should be released from the lee bow so as to spread some distance to leeward besides providing a windward slick. In past years a suitable method of providing the slick was through the utilisation of forward W.C. pans, but now that the forecastles of ships only provide stowage space, other methods are indicated. A very convenient method consists of securing a length of water hose to the lee bow guard rail stanchions, sealing one end and turning the other up to the upper rail. This length of hose can then be pricked at intervals of about three feet. When this hose is filled with oil at the turned up end, the oil will run out of the holes and down the bow plating to spread effectively both to windward and to leeward. (Figure No. 52).

When towing in heavy weather, the effect of oil is most marked; when the tow is structurally sound and whole the use of oil is advantageous, *but its effect when the tug's charge has suffered structural damage must be seen to be believed.*

If the tow is following well without undue yawing then the best means of distributing oil is from forward in the tug on both sides, using forward waste pipes from either washrooms or toilets, by this means both tug and tow derive advantage from the oil. It will be apparent however that the

172

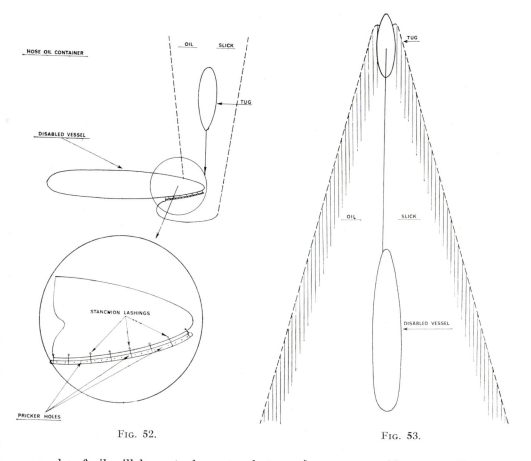

FIG. 52.   FIG. 53.

supply of oil will have to be somewhat on the generous side to provide an unbroken slick right back to the tow.   (Figure No. 53).

In the event that the tow is yawing to any extent the effective distribution of the oil becomes complicated.   It is quite common to experience yaw in a tow extending over eight points, or more, of arc . . . viz. four points on each side of the course made good.   If the towing medium in use approximates to 250 to 300 fathoms of gear then no single oil slick laid by the tug can hope to benefit the tow for any more than a small percentage of the time.

A practical solution to this problem is provided by the construction of a heavy canvas bag of dimensions approximately 36 in. × 18 in. (diameter) arranged to taper to one end where, after the bag has been stuffed with cotton waste, a screw bottle neck, broken off from a pint beer flagon, can be sewn in. This bag should be provided with beckets and a large bow shackle for securing to the towing wire. The bag should be pierced in its lower half by numerous holes. It may be necessary to ballast this bag in order to ensure that it will assume an upright attitude below the towing wire when streamed. (Figure No. 54). The bag should then be filled with

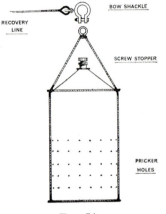

RECOVERY
LINE

BOW SHACKLE

SCREW STOPPER

PRICKER
HOLES

Fig. 54.

oil and streamed astern on a small wire retaining line. Depending upon the speed of advance of the towage unit and the depth of the catenary in the towing medium and, of course, the presence or otherwise of joining shackles, the bag will assume a position a little more than half way along the gear with the oil coming to the surface somewhere farther astern than this, in any case, such as to provide a more effective slick than by any other means available. (Figure No. 55).

A bag of the dimensions indicated is effective for from 9 to 12 hours. Vegetable oil is the most satisfactory oil to use for the purpose but diesel oil and furnace oil are effective in use if stocks of vegetable oil are not equal to the demands of protracted bad weather. It is probably quite superfluous to observe that such a bag will require the assistance of the capstan when

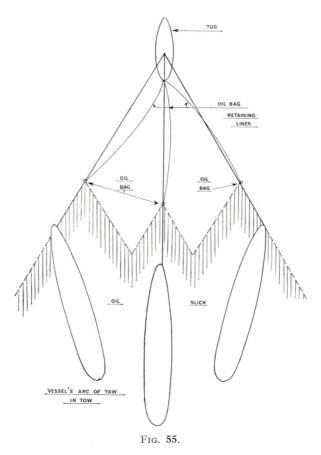

FIG. 55.

replenishment is indicated, also that there is always a fair chance that it will foul a shackle and become unrecoverable in any case. It is not super-fluous to remark that even if two or three bags have to be constructed during a spell of adverse weather, the excellent effect deriving is amply compensatory for the time and trouble expended.

CHAPTER X.

Open Seas Boatwork in Association with Ocean Towage Operations.

1. General.

The launching of boats from ships at sea under favourable conditions of weather presents no difficulties which are not readily resolved by reasonable standards of seamanlike proficiency and boats, and associated equipment, provided and maintained at the averagely excellent standards which prevail in the Merchant Navies of the world generally. Under conditions of adverse weather, however, such operations become a matter for considerable seaman-like skill and the most accurate appreciations of all of the various conditions obtaining; boat-work in bad weather calls for the absolute in co-operation and teamwork from all concerned, the Commanding Officer in his overall command of operations and his manoeuvring of the parent vessel, the inboard hands in their handling of the winches, falls, painters and other impedimenta, and last but certainly not the least, the boat's crew proper who will require a deal of courage besides the more obvious qualities which are indicated. The decision to launch a boat under circumstances of adverse weather is therefore not one to be lightly implemented because, by the very nature of operations, there is ordinarily no turning back once the boat begins its downward passage from the davit head to sea level. In all of the relevant considerations a Commanding Officer must bear in mind not only the problems involved in safely launching the boat, but also the more difficult conditions which must certainly attend upon its recovery, when the boat's crew may well be tired to the point of exhaustion, with all of the added risk to life, limb and property which is thereby entailed. Whilst a Commanding Officer must, under extreme circumstances, accept the non-recovery of the boat he cannot, under any circumstances, contemplate the loss of the boat's crew.

In almost all seagoing vessels there are physical features and operational considerations which make for difficulty in the matter of boatwork and boat launching and in this the ocean-going tug type is no exception. In tugs the principal difficulties derive from the relatively small bulk of the type which tends to offer a very lively platform for any form of open seas working. This tendency is not in any wise ameliorated by the presence, in most tugs, of a massive all-round timber or steel belting which provides a substantial obstruction to the passage of boats upwards and downwards from the davits and also adds to the difficulties of alongside working. The means and procedures adopted to overcome these, and other and allied practical complications, under bad weather working conditions forms the principal burden of that which follows.

Before proceeding with the detail of this chapter it is perhaps permissible to remark, at this stage, that there are a number of sea callings, other than sea towage, where the function of a vessel extends over into open seas boat employment, examples are:—Sea Station Pilot Cutters, Lighthouse and Lightship Tenders, Cable Ships, Salvage Vessels, Surveying Ships and certain varieties of Deep Sea Fishing Vessels. It is no casual coincidence that the procedures, equipment, modifications and usages which will be mentioned here are repetitive of well established customs in these vessels.

2. Boats, Boat Handling Gear, and Boat Equipment.

It has often been remarked that whilst all of the sea service ship types which were mentioned in the last paragraph of the preamble above, have developed specialist boats for their own particular needs this does not appear to have been the case with Ocean-Going Tugs. This is, of course, quite true but this circumstance does not result from any parsimony on the part of the Operators, or from any lack of initiative on the part of the seagoing personnel, as may appear; it results from a lack of suitable space caused by the relatively small size of tugs, in the first place, aggravated by prime design requirements in the second, which has left the barest possible space, even in the largest classes of tug, for the proper accommodation of the statutory equipment of lifeboats only. It thus became the custom to effect such modifications as were permissible to this equipment to serve operational needs. Such modification has usually taken the form of providing a motor for one lifeboat, the chosen boat being built to the transomed shape with some vestigial decking forward and aft. Notwithstanding this very mild modification, tug's boat equipment was, for many years, primarily statutory both in form and in function so that in consideration of past endeavours by ocean-going tug personnel in boats in the open seas, one cannot but admire their courage, skill and tenacity.

Recent official approval for such items of life saving equipment as the inflatable liferaft has now, however, rendered up space hitherto occupied by lifeboats for the accommodation of craft to a functional design and purpose so that the more recent tonnage introduced into ocean towage service has been seen to have been provided with two, and sometimes three, boats designed expressly for the service. One item of this new style equipment consists of a boarding boat specifically designed for personnel transfer at sea and which has been based upon the boarding boat type which has been evolved for the use of the Sea Pilot Cutters which serve the offshore stations. Such boats remove much of the hazard of boarding operations because of their strength, seaworthiness provided by an enclosed design, and their specially devised boarding and landing areas surfaced with non-skid material and contained within properly devised guard railing. These boats, usually built to a length of about eighteen feet, with other dimensions to proportion, are the product of a half a century of davit and alongside working experience; they happily accommodate a working complement of twelve passengers so that they are wholly adequate to all boarding requirements and, because, it is the custom to fill every suitable space within them with buoyant material, they inspire confidence in the personnel which they were conceived to serve.

A second boat, of the same general characteristics, but to a length of twenty-six to twenty-eight feet, is becoming a part of the new pattern also. These boats are intended for the transport of stores, spares, salvage pumps and hoses, working parties and the like, the necessary flexibility being obtained by making certain components such as thwarts, side benches, sprayhoods and deck sections portable. Arrangements are also commonly included in these boats for the running off of hauling lines, towing media, and for the transport, upon occasion, of anchors of modest proportions.

These two boats are essentially rugged work boats and, as such, will ordinarily be powered to provide a service speed of about eight knots by means of single diesel installations. When a third boat is carried it is usually a lighter and faster craft, often powered by an outboard motor for maintenance convenience, and which will provide a ship to shore communications link whilst on salvage station. Boat construction in Glass Re-Inforced Plastics is becoming increasingly acceptable to ocean tug operators, not only because of its immunity to decay and its resistance to shock and impact damage, but because of the practicability of 'on the spot' repairs by tug personnel. All boats for tug use must, however, be provided with substantial all around fendering regardless of their construction because of the rigours of the service. This fendering, which should be of 'D' sectioned rubber of substantial size, must be effectively secured to the sheer strake close up to the gunwales and provided at double thickness at the stem and about the quarters of transomed boats.

The use of independent electric boat winches for each pair of davits and an effective quick release device in each boat is quite mandatory in tug service. The complex and serious stresses which are suffered by all of the separate components of sea boat working gear are very much reduced if an efficient shock absorbing device can be incorporated into each boat fall. This feature can be worked either into the davit equipment or form part of the gear inboard of the boat, both are effective but the latter is the better.

Most well run ocean tugs augment the major items of boat handling equipment with certain well established modifications and associated practices intended to enhance the safety and efficiency of boat conduct both inboard and afloat. The effect of a lively tug's movement upon a boat freely suspended in davits will be entirely obvious to all; it will also be quite apparent that the effects of both roll and pitch must increase in proportion to the scope of fall veered to any boat in davits, so that the movement of any boat, in response to the parent vessel's movement in the seaway, must become increasingly violent as she descends towards to sea level so as to become virtually uncontrollable and liable to serious structural damage, not to mention the hazard to which the occupants of the boat will be subject. This condition is contained, in ocean-going tugs, by rigging a pair of bottom lines to shackle up to the davit heads in use on one side and set up to a pair of handy billy tackles on the other. When the davit head is either wound out, or run out according to type, these bottom wires are set up taut by the tackles provided. The boat is then bowsed in to these bottom lines by steel wire rope lizards passed around them from the bow and stern and turned up inboard. By this means the boat is held pinned in to the tug's side whilst launching and recovering to the advantage of all.

It is the common practice also to rig light tripping lines to the lower terminals of boat falls through small blocks shackled up to eyes provided at the davit heads. By means of these lines the falls are swiftly triced up clear of the boats and their occupants the moment that they are slipped, to be held at the correct length and in full readiness for a speedy reconnection when the boats return alongside.

Whilst all tug's boats must be properly equipped with a painter for orthodox usage and for sea anchor application, this item is not used for working alongside of the tug; instead, for convenience and efficiency, inboard painters are passed outboard from the tug. The forward painter, usually provided in duplicate in case of mishandling, is made up to such a length as to extend aftwards from their forward securing point in the tug as to allow of working on the towing deck as well as under the davits, the length being such as to allow the end to be brought inboard, with the boat being made fast upon the bight. In bygone years these painters were commonly made up from grass fibre ropes because of their properties of buoyancy and extensibility; modern usage, however, trends towards the employment of either Polythene or Polypropylene man-made filament ropes for the same reasons but interpreted to greater efficiency. To preserve these painters against inboard damage they are commonly made up, at their forward ends, to steel wire rope snotters of appropriate dimensions to accept the chafe punishment which must arise over gunwale detail and at the securing points. It is also customary to mark the painters, with red bunting worked into the lay, at positions appropriate to the under davit and after working deck locations for the guidance of the bowmen.

Stern painters are similarly arranged except that they are set up inboard of the boat to a ring and toggle for rapid slipping and are tended, and made fast and cast off, as occasion demands by an after painter hand stationed suitably aboard of the tug.

Because the belting feature of ocean tugs is provocative of very great inconvenience to boat working, and not infrequent positive danger, certain operators have elected to dispense with it about the mid sections of their tugs, retaining it only about the bow and quarter areas. Although there is a great deal to recommend this action it does not commend itself to all so that those who elect to retain full length belting should be encouraged to fit fairing detail both above and below the belting, not only beneath the davits but also at a suitable location abreast of the towing deck.

3. Maintenance.

Seeing that ocean tugs' seaboats, together with their equipment and handling gear, may well provide the most critical factor in operations of considerable financial significance, their maintenance to high standards is clearly quite mandatory. Boats which must stand in chocks for weeks on end and in climatic conditions which may vary from the tropical to subarctic must clearly be vulnerable to material and mechanical deterioration. It is not then nearly sufficient to have the ship's carpenter wave his oil can in their general direction at weekly intervals, and a rigid schedule of maintenance drill is essential. Engines must be run regularly and electrical equipment must be tested, moisture absorbent packs being distributed to best advantage. Boat stores and equipment also require regular checking in view of their potential importance. In tugs, as with all small sea service vessels, it is the well-proven case that a sound, well-fitted boat cover is second only in importance to a whole hull.

Inboard gear such as davit mechanisms and winch motors must be worked regularly and serviced properly if untoward defects are not to arise at critical moments. Sheaves, blocks and leads require greasing and moving whilst all of the running gear, falls, tripping lines, painters and bottom lines must be thorough-footed from time to time and turned end for end to schedule.

Because tugs are obliged to occupy salvage stations at distant places, frequently at anchor, ship's personnel are enabled to reach high standards of boat handling proficiency as a part of their day to day duties. The occupation of exposed anchorages also provides the opportunity of practising boat launching and recovery under conditions which closely resemble the operational, thus affording a rotation of duty, inboard and outboard, which is quite invaluable. In every aspect of maintenance, whether it be in terms of mechanical, material or personal efficiency, sight should never be lost of the fact that open seas boat work is a dangerous business from many more points of view than the simple prospects of immersion.

4. Launching and Recovery Procedures.

The practicability of boat working having been established to the Commanding Officer's satisfaction, and proceeding upon the assumption that the purpose of launching a boat is to place a boarding party on board of an abandoned casualty in order to make a towage connection, the very first requirement must be to ensure that the boarding party is adequate to the occasion. The strength and composition of the party will largely depend upon the manning of the tug but must, in any case, provide a sufficiency of strength to manhandle enough towing wire rope inboard of the casualty to effect a connection according to the circumstances. Most ocean-going tugs can afford a boarding party comprising five ratings, three of whom should be seamen ratings. If an Engineer can be spared to accompany the party, this is strongly recommended, because, especially in modern ships where auxiliary machinery is independently powered, an engineer of initiative can occasionally make a substantial contribution to the health of an operation by obtain-

N

ing power.   Boarding parties are traditionally led by the Chief Officer, a custom which has everything to commend it seeing that so much will depend upon the initiative, energy and resource of the leader at every stage of any boarding operation.

Practical preparations for boarding will commence with the boat with the insertion of the bottom plugs, a fact which must be reported to the Commanding Officer.   The boat engine will then be started and run up to working temperature, an Engineer, preferably the boarding engineer, being present through this phase.   Engine controls will be checked and worked. Meanwhile the bottom lines and the painters will be rigged and all boat stores and equipment will be checked.   There has always been a deal of controversy over boat equipment and the complaint is often uttered that there is no space in the boat for movement because of a super-abundancy of boat gear.   It is perhaps, however, significant that this complaint is never uttered by personnel who have been, for any one of a number of reasons, constrained to remain in a sea boat for any substantial period.   The prudent view in this matter is that no boat should leave a parent vessel short of any equipment that can possibly serve the boat party in any forseeable contingency.   If VHFRT equipment is carried . . . (and this equipment is quite invaluable) . . . a test exchange should be made.   Finally the fuel tank should be topped up.

The boarding party should be encouraged to change into clean, dry and warm clothing.   They should be provided with life-jackets, which should be fitted and secured under the eye of the Chief Officer, and a pair of working gloves to protect hands during boarding and hauling in procedures.   When boat and crew are in full readiness this should be reported to the Commanding Officer who will then consider his approaches to the vessel to be boarded.

Reference to Section IV of this work will show that vessels lying stopped in the sea will assume angles of rest and will drift according to wind direction and the amount and disposition of their exposed and immersed areas.   From this it follows that all vessels, under these conditions, must offer upon the side away from the wind, an area of lee which will proportionate to these factors.   This area will obtain at its most favourable in the case of a large vessel lying beam on, and at its worst in the case of a small one lying head on or stern on to the wind, with all of the endless variety which can clearly present with intermediate cases.   *This will provide the key factor in any Commanding Officer's assessment of the practicabilities inherent in any given situation.*   Action as recommended in Section IV, Chapter 2 will provide an estimate as to the rate of lee drift. The area of lee deriving from casualties is triangular in form and of a precise shape depending upon the factors described, but it usually extends to about one and a half times the casualty's length downwind and is sufficient to screen a tug from wind and sea, if not from swell effect, over the period required for a trained and practiced crew to launch a boat.

With the appropriate assessments made the Commanding Officer will order the boat launching party and the boat's crew to stand to stations and will then take the tug down to leeward of the casualty, passing under her bow or her stern according to her attitude and lee drift, to assume a position down-wind of the apex of the triangular area of lee. As the apex of the lee area passes over the tug appropriate helm and a touch of ahead engines should be used to incline the tug away from the fore and aft line of the casualty so as to facilitate a rapid departure from the situation once free of the boat. The Commanding Officer will then order the boat away.

The actual initiative in the expedition of affairs passes, at this stage, to the Chief Officer who will have embarked his crew into the boat and will be in process of winding out. As the weight comes on to the falls he will order the lizards to be passed and turned up to hand forward and aft and the painter hands to turn up and be at readiness to render. He will then order the lowering of the boat, adjusting the rate and ordering a halt as he assesses the period of the swell and sea because his aim must be to drop his boat into rising water. Satisfactory conditions being achieved he will order a lower away and will slip the falls so as to drop the boat a foot or two from the sea surface. He will simultaneously cast off the lizards and the after painter. The boat will have been lowered with the engine running and with all hands in the seated position with one hand functionally employed and with the other holding on. At the moment that the boat is waterborne, the Officer in Charge will come ahead on the boat engine with the rudder away from the tug to sheer clear on the painter which will be cast off when clear to be hauled in by the inboard painter hand.

The question of whether or not the boat used shall be made the lee or the weather boat depends greatly upon the lee presented by the casualty. If the lee is effective then the terms 'Lee' and 'Weather' become largely academic in their application. Where the conditions of an operation fail to reduce the lee drift of the tug, the prime problem is to avoid the trapping of the boat under the tug's belting, a condition which can become most hazardous if the tug is allowed to roll heavily. In such cases it has been found advantageous to promote slight headway on the tug so as to ensure a rapid response to the helm and to drop the boat at the belting level. Other Commanding Officers incline towards the practice of making the working boat the weather boat under these conditions.

The whole operation of dropping a boat into the lee of a casualty requires a great nicety of judgement because any form of close working obliges an acceptance of collision risk. Because practical experience in this operation is rare, one Tug Commanding Officer of considerable renown spent a week or more on board of a cruising pilot vessel to assimilate and observe the practice of experts in the field. A close approach is, however, entirely essential, not only to allow of the advantage of the casualty's lee for improved performance, but so as to embark the boarding party into the casualty with the minimum of effort and exposure.

Actual embarking processes vary so greatly as to preclude any endeavours towards description, suffice it is perhaps to remark that access to such vessels is often provided via falls and the like trailing overside after the departure of boats. Grapnels upon heaving lines occasionally provide a method of ingress whilst, upon at least two occasions, boarding has been facilitated by the tug proceeding to windward to fire rocket lines clear over the ship. It follows that at least one member of a boarding party should possess an agility equal to swarming up a single rope to a casualty's deck to provide a fairer access for the remainder.

Retrieving a boarding party after a successful connection has been made is invariably a more difficult operation principally because the lee from the casualty is no longer available. Whilst instances are on record of Commanding Officers who have found it possible to hold a position in the lee of their potential tows whilst recovering and restowing their boats, it is clearly only the very expert Captain who is able to encompass the very close quarters manoeuvring that is entailed whilst hampered by a towage connection. Ordinarily the attempts at recovery will be made when the towing medium is fleeted and when towing is in process.

When the tug is ready to take the boat back inboard, the Commanding Officer will assume a heading which allows of the steadiest possible platform and will reduce speed to bare steerage way. The inboard hands will then rig a scrambling net over the bulwarks at the chosen after working position, two hands should go outboard upon the belting upon man-ropes, and the fore and after painter hands should stand by with the ends of their respective lines made up to a heaving coil. The boat should then be called in, the painters passed and set up, and all of the boarding party taken out with the exception of a coxswain and a bowman. Boarding party members should leave the boat for the net on rising water assisted by the hands working outboard. Obviously this phase of the operation will be carried out as quickly as may be arranged.

The boat should then be sheered off under her own helm and engines, the two inboard painter hands ensuring that no bights fall into the sea to risk fouling the boat's propeller. The davits are then wound or run out and the boat is brought forward to allow the bowman to turn up the forward painter at the appropriate mark whereupon the stern painter hand takes a turn inboard of the tug and sets up aft. The falls are then sent down on the tripping lines to be engaged forward and aft, in that order, by the bowman and coxswain respectively. As soon as possible thereafter these two ratings pass the lizards about the bottom lines and then sit down in the boat and hold on. The order to heave up is then given and proceeds with all speed until the boat is up to the davit heads and can then be restowed and secured.

The hazards of recovery and restowing will not be laboured here, everything will depend upon the proficiency of the officers and ratings concerned. Perhaps it is permissible to observe that, with the weight of towing, the Commanding Officer is enabled to apply power for rapid directional changes of brief duration such as may serve to counteract the lively movement which is a feature of all tug seaway working. Perhaps, however, the most significant observation which can be offered in this context is that boat winch motors should be of such power as to be able to reduce the hoisting time to the absolute minimum so as to reduce the time spent by the boat in the 'between wind and water' condition and where most of the damage and hazard obtains.

The use of oil whilst carrying out boat operations is somewhat of a mixed blessing. Whilst it cannot be denied that the use of oil does improve sea conditions, it must be observed that one has only to ship a very few gallons of oily sea water into a boat to make any subsequent movement therein difficult.

CHAPTER XI.

The Conduct of the Tow.

The overall length of a towage unit comprising a tug or tugs, and the vessel or object towed, is rarely less than 1,000 ft. in length and is often well over 2,000 ft., its speed of advance only infrequently exceeds five knots. Whilst the tug or tugs involved constantly endeavour to maintain accurate courses, the vessels and objects towed, for a variety of reasons, adopt an erratic and serpentine course astern of them. It follows therefore that such units are much more vulnerable than most to the hazards of navigation and traffic.

Towage units are not commonly met with on the high seas so that the Watchkeeping Officers of Ocean-Going Tugs must always be fully cognisant, not only of their charge's special vulnerability, but that the majority of the Watchkeeping Officers of other ships encountered will not have a conditioned reflex to the appearance of a towage unit. They will not usually have any but the vaguest idea as to the capabilities of the unit in terms of either performance or manoeuvring. It will also be doubtful if they will recognise the true line of advance of a unit if this is at all complicated by a yawing tow. It is also entirely probable that they will not appreciate the effect, upon a towage unit, of a current or tidal set if this is not unduly affecting their own ship. This is offered at some length because the International Rules and Regulations for Preventing Collision at Sea do not provide any special recognition of the difficulties of a towage unit. Whilst it is true that Watchkeeping Officers in ships of most flags will give way to a tug and her tow, this is not obligatory upon them and it is probably all the more hazard provocative that this courtesy is sometimes extended and sometimes withheld. Within the context of the traditional expression of 'Fog, Mist, Falling Snow or Heavy Rain' . . . and when Radar is in use, all of the foregoing gains added significance.

It is beholden upon the Watchkeeping Officers of tugs, in view of the special vulnerability of their composite charge, to meticulously obey each and every requirement laid upon them by International Regulations in the matter of lights and sound signals. It is entirely essential also that both the letter and the spirit of Article 4(a) of the International Rules should be most faithfully obeyed so that 'Not under Command' signals are not displayed unless the unit is *in fact* . . . 'unable to manoeuvre as required by these rules' . . . .

If the vessel or object towed is manned then constant and reliable communication must be arranged and maintained between tug and tow. This may be arranged by V/S or by VHFRT. The former is reasonably efficient providing daylight signalling lamps are available and maintainable at both ends of the tow. The latter, under circumstances allowing of proper maintenance, is however infinitely more convenient. The Towing Appendix of the International Code of Signals is a somewhat limited and rather laborious method of exchanging information and advice, but it must not be condemned out of hand because it provides an effective, if limited, emergency provision.

Communication between tug and tow must be organised upon a periodic reporting basis, besides the normal service exchanges, in order that untoward occurrence may be provided for.

The prime concern of the Commanding Officers, Watchkeeping Officers and Ratings of both the tug and the vessel or other object towed must clearly be the preservation of efficiency in the towing medium. Aboard the tug this necessitates a constant watch on the towing deck or in the towing winch room; winch settings and the scope of gear in use must be carefully related to the conditions of weather, sea and swell prevailing and the behaviour of the tow, adjusted from time to time according to standing orders interpreted by the Officer of the Watch. Concerning this attention is drawn to a practice in certain well-ordered tugs of keeping a Winch Room Log where all relevant data is entered at regular intervals.

All points of contact between inboard fittings and the towing gear must be effectively lubricated and such anti-chafe measures as may be necessitated by the type of equipment in use must be maintained to a high standard. The Officer of the Watch should personally inspect all towing arrangements *before* and *after* assuming responsibility for the Watch, not neglecting to examine the towing bridle to ensure that it is functioning according to requirements.

When the vessel or object towed carries any part of the towing medium inboard then towing must be eased at regular intervals to allow the nip in inboard gear to be freshened. This practice is quite essential because where cable links bear heavily upon details of windlass, cable compressors or hawse pipes, fatigue can develop readily. When the inboard sections are of wire rope then this pause can be utilised to renovate chafing arrangements besides adjusting length. Where wire is in use the utmost vigilance must be maintained in view of the severe crushing and torsional stresses endured because failure sometimes occurs at very short notice. It is a common practice on long tows to ease towing at noon of each day when the signal to resume way is made by the tow and is accepted as an indication of satisfactory conditions at the towed end.

Should auxiliary power be available in the tow for the purpose of steering then all towing stresses can be substantially reduced by the maintenance of fair station astern of the tug. It must be appreciated however that steering a towed vessel is not a responsibility which is easily discharged because, besides the relatively slow speed of tow through the water, there is a tendency in most towed vessels towards yawing to a greater or lesser degree besides which the weight, and the checking propensities of the towing medium, causes an insensitivity to more normal angles of helm so that Quartermasters are obliged to keep control through large and frequent applications of helm. This requires great concentration and vigilance with all of the strain involved. Steering by compass, whilst in tow, is rarely practicable and best practice is to steer by visual quartermaster con upon the tug, an upper sternlight of suitable intensity being displayed by the tug during the hours of darkness to facilitate matters.

Before commencing any towage operation due consideration must be given to all available weather, tidal or current atlases, or other nautical publications descriptive of the areas to be traversed. Advantage may well be taken of this preparative phase to effect this review in close co-operation with the officer commanding the vessel, or object, to be towed, this in itself providing some intimation of the collaboration and operational liaison which a successful towage operation demands.

Because of the slow speed, limited manoeuvrability and general vulnerability of towage units on passage, every advantage must be taken of such natural accelerations as may exist in terms of wind, tide or current both seasonal and permanent. Whilst the greatest advantages from these will clearly obtain upon the ocean stages of the operation, the morale of the ships' companies of both tug and tow will profit from a proper consideration of tidal conditions when decided advantage may derive from a selection of favourable tidal phase when commencing, or completing, an operation or when traversing narrow waters where dense traffic prevails.

Successful coastwise and narrow seas navigation entails the most accurate allowances for the set and drift of the unit from all causes, estimations which must inevitably seem disproportionately large because of the low speeds achieved. In this consideration the attitude of the unit to its true line of advance must always be borne in mind, particularly in the terms of impinge-ment upon . . . 'the other ship's water' . . . in the narrower channels, and its visual impact upon approaching traffic. The necessity for tugs to over-run the more normal alteration of course focii in order to keep their tows within channel limits, and clear of channel markers, must also be remembered in the same context.

Besides the basic depth of water considerations which are obligatory upon all navigation, those concerned with the safe conduct of a towage unit must always bear in mind the depth of the catenary in the towing medium. They must also give thought to the added hazards provided by sea-bed fouled areas at depths considerably in excess of those normally associated with surface navigation.

The ocean navigation of towage units profits substantially from a close liaison between tug and tow when fix and position line data, together with resultant course and speed determinations, can be passed from one to the other to great mutual advantage. Co-operation of this kind is quite invalu-able to the tug Commanding Officer, particularly in foul weather, when the relatively easier movement and more elevated height of eye obtaining in a large tow, not only permits of more convenient solar/stellar observations but affords greater protection to the observer, and his sextant, against seawater and spray. Celestial observation from tugs in adverse weather is never easy due to the lively movement of such craft in a seaway, but when heavy swell forms part of a foul weather condition, the tug's horizon is as often obscured as not. It is also a point for worthwhile consideration that patent logs, whether of the impellor or the streamed rotator types, return more reliable information when operated from the less lively hulls.

The state of the weather will, of course, always provide the major pre-occupation of all concerned with ocean towage operations and it has long been the practice for Ocean Tug Navigators and Radio Officers to co-operate in taking the fullest advantage from the weather reports and forecasts which are disseminated by the Meteorological Services of the world's maritime nations. The compilation of weather charts for the areas through which navigation proceeds is a standard practice which provides an ample return where speeds of advance are commonly very low because, by this means, critical areas may be avoided and more favourable conditions enjoyed. In this context the following observations are offered upon the conduct of tows when weather of such severity as to necessitate a consideration of heaving to threatens.

The normal benefits of heaving to, in terms of eased conditions, do not accrue to tugs and their tows to the same degree and effect as they do to single vessels because the engine power required to be exerted by the tug in order to maintain the orthodox hove to attitude so frequently approximates to full power as to afford little or no relief of stress to the tow-rope and its associated components. As a result of this, Tug Commanding Officers faced with such conditions, and where there is a sufficiency of sea-room, prefer to turn and run with the weather astern rather than to attempt heaving to; although clearly the decision to take this action is not one to be unduly delayed. With the tug and tow running before the weather, speed may be reduced to bare steerage way because the tug, by virtue of her 'stern up into the wind' propensities, will hold to her course steadily, a tendency which will be assisted by the weight of towing although firm bridling and a continuous watch upon the gear will become a vital essentiality. Speed before the wind will naturally vary very considerably according to the physical characteristics of the vessel or object under towage in view of the risk of broaching to but collective experience in such exercises would appear to indicate that the inter-reaction of tug and tow, together with the weight and resistance of the towing medium in use, tend to produce very reasonable course maintenance.

Notwithstanding the fact that all and every seamanlike precaution has been taken and every professional obligation faithfully discharged, vessels and other floating objects under towage, do founder from time to time. Whilst units under towage have been known to sink rapidly and unexpectedly because of damage of a nature and to an extent unappreciated by those in charge, this is a circumstance related more to the perils of sea warfare than the accidents of peace time voyaging and the more common circumstance is for them to founder for reasons apparent to all. The normal processes of foundering are, in fact, in all but the smaller classes of vessels, relatively leisurely so that there is usually time to effect all of the dispositions which are necessary. In every case of such risk, however, it is essential to remove the crew of the threatened unit back on board of the tug before the onset of actual hazard, and all other activities must necessarily be conditioned by this prime obligation.

If the sinking takes place in deep water the sacrifice of the towing medium is obliged upon the tug if damage and possible serious risk of capsizing is not to result but in shallow water it entirely is practicable for the tug to recover a substantial proportion of the gear in use. This is readily achieved by allowing the tug to drop to leeward of the position of the foundered tow upon the towing gear well bridled down aft. When the tug has 'brought up' as it were, the gear may be retrieved by the ordinary processes of shortening in accordance to the equipment available.

Before finally severing its connection with a sunken tow it is incumbent upon the tug to accurately establish its position by the best means to hand. This information, together with the depth of water at the position and the depth of clear water over the hulk, must then be transmitted by the most rapid means to hand to the appropriate Hydrographic Authority. The position should also be effectively marked and for this purpose well-found tugs carry dan buoys painted green and bearing large green flags, together with sinkers and riser materials of appropriate sizes.

When the seaworthiness of a tow is in question the Tug Commanding Officer concerned must order his navigation so as to incur the minimum risk of fouling recognised shipping routes, channels and anchorages or roadsteads. He must moreover never attempt an entrance into, or passage through, the precincts or any Port or Harbour without informing the Harbour Master, Port Captain or other equivalent authority of the fullest details of his charge

CHAPTER XII.

TUG COMMUNICATIONS.

Efficient communications arrangements, internally, ship to ship and ship to shore, are possibly the most important factor in meeting the demands for increased productivity and economy in modern maritime activities. In no aspect of such activities is this need more evident than in the sphere of ocean towage, whether in the salvage/rescue operations or in the more prosaic proceedings associated with contract towage. Indeed the first mentioned function imposes an additional, and quite unique, demand in terms of the truly continuous watch which the occupation of a salvage station requires and which, in certain circumstances, may involve the continuous operation of three separate types of equipment.

The scope of towage communications may, for convenience, be summarised under four broad categories:

1. Internal communications facilities.
2. Communications equipment covering the visual ranges.
3. Short range coverage by Very High Frequency Telephonic Equipment.
4. Medium and Long Range Radio/Telephony and Radio Telegraphy competence.

1. Internal Communications.

Whilst it is true that every ship afloat requires a properly integrated intercommunications system, this demand becomes somewhat intensified in those types of vessels which are commonly designated as Service types. Vessels, that is, intended for use as a tool rather than those which may be broadly described as carriers.

In vessels where both helm and engines are in more than averagely frequent movement, and where machinery or appliances, other than main propulsion machinery, all in operation and where the vessel's function extends outwith her own fabric, sometimes in conditions provocative of hazard, there is a very pressing demand for fully effective intercommunications.

This is particularly the case when the responsible personnel, at their various positions of control, are so actively engaged that they cannot leave their positions in order to communicate.

In the ocean-going tug type the necessary facilities are provided by an internal telephone system connecting selected stations by means of a dial switching system. Where the stations are greatly affected by machinery or other noises, best practice provides klaxon calls, amplifiers and accoustic booths. Ordinary seamanlike prudence dictates the necessity of locating the separate stations with care and with a consideration for conditions as they obtain in active towage service. An emergency arrangement of large bore voice-pipes between the more important stations is strongly recommended.

2. Visual Range Communications.

Indulgence is requested for beginning this part of the discussion with observations upon Limited Range Very High Frequency Telephony as represented by private wavelength portable instruments with a range of 1,000 metres and under. *Such equipment has no equal in the discharge of communications responsibilities over the ranges commonly associated with towage operations.* Indeed, once the convenience and efficiency of this sort of equipment has been experienced, particularly upon operations of the more complicated kind, there is a marked reluctance, upon the part of all concerned with the practical aspects of operations, to revert to older and more laborious processes.

When V.H.F.R.T. equipment is not available, recourse is ordinarily made to communicating by means of daylight signalling lamps and most modern tugs are very well favoured with such equipment, many being provided with a pair of exterior shutter type lamps located upon mast complexes, or elsewhere as convenience allows at a good height, and remotely operated from the wheelhouse. In this connection it must be observed that whilst the Semaphore System appears to be dying a natural death in other aspects of sea employment, it still endears itself to tug personnel, particularly when engaged upon multiple operations, because of the speed and convenience which is allowed by the rate of one letter per gesture with equipment which costs but pennies only and offers no maintenance worries.

Mention must be made here of the towage groups in the International Code of Signals, Volume I, *Remne.* Although the facility thus provided is somewhat limited in its application, tedious in operation and, in certain instances, obsolescent in application, it has the great merit of being internationally comprehended. Because of this it is incumbent upon all tugs to have two pairs of halliards available for the aftwards display of bunting.

3. Very High Frequency Radio Telephony.

The universal acceptance and application of this gear renders its inclusion into towage practice quite mandatory. Besides its obvious value in terms of instantaneous, procedure free, plain-language communication, its value to tugs entering, leaving or traversing Port Areas possessing Navigation or Communications Centres may only be extravagantly expressed.

This particularly in so far as it allows a Tug Commanding Officer the facility of advertising, to all concerned, such inhibitions to free progress and manoeuvre as may apply to the unit in his charge. Hardly less important is the pre-consultation which is so conveniently allowed with shore interests by this means, not to mention the readiness with which towage arrangements may be discussed with vessels, or other floating objects, requiring salvage or rescue services.

4. Medium and Long-Range Wireless Facilities.

It is probable that the ocean-going tug type was among the earliest and most enthusiastic converts to radio usages, if only because of the added professional efficiency deriving from the rapid reception, or interception, of news regarding casualties at sea.

The importance which still attaches to radio-communications in tugs is amply demonstrated by the comprehensive range of radio equipment which is mounted in tugs and the extent to which tug operators are prepared to go in the employment of staff.

Radio Offices in modern tugs are of equal importance with both wheelhouse and manoeuvring platforms and are located and furnished accordingly. The best practice is to locate the Radio Office immediately abaft of the wheelhouse with a communicating door or other aperture. Because of the necessity for occasional direct communication between the Commanding Officer of a Tug and the Master or other Officer in Charge of any vessel or other floating object requiring assistance, or in process of receiving assistance, it is the growing practice to arrange remote extensions from both of the Radio Telephony Equipments in positions conveniently disposed in the wheelhouse.

SECTION 5.

CHAPTER I.

The Role of the Ocean-Going Tug in the Refloatation of Stranded or Grounded Vessels.

It is necessary at the very outset of this section to sharply differentiate between the Ocean-Going Tug and the Salvage Vessel. Ocean-Going Tugs are all too commonly designated as Salvage Tugs and it must be very clearly established that the latter appellation can only truly apply when a vessel combines the attributes of both a Tug and a Salvage Vessel, with all of the personnel and equipment necessary to the dual function. That such vessels do exist cannot be denied, but it would seem that, like most hybrids, such vessels must fall short in both functions seeing that certain qualities essential to the Ocean Tug are the precise antithesis of some necessary to salvage work and vice versa.

Without digressing too far into the design of the two types, certain examples in support of the contention must be offered. For instance, the very features which permit ocean tugs to proceed to sea at speed in adverse weather, viz. the high forecastle heads incorporating widely flared bows, their enclosed superstructures and deep draught are in pure contradiction to the characteristics necessary to salvage work. Salvage Vessels spend a great deal of time on refloatation work so that modest draught is quite essential. To permit of efficient attendance upon divers, camels, elaborate and extensive ground tackle and to facilitate overside construction and repair work, a flush unhampered deck is an imperative need; whilst a high and amply flared bow could be nothing but an embarrassment to a craft which is compelled, by the very nature of her function, to manoeuvre among, and to lay alongside of wrecks, derelicts and casualties in all weathers and, as often as not, in open waters.

In the field of groundings, strandings and beach work in general, it is fairly analagous to compare the ocean tug to the Casualty Officer of a large hospital. Both are poised, ready at a moment's notice, to proceed to the scene of an accident to render immediate aid. In just the same way as the

Casualty Officer is prepared to deal expertly and expeditiously with a straightforward fracture or laceration, leaving the treatment of the more involved and complicated injuries to the full facilities of the hospital, so is the tug competent to deal with a simple stranding or grounding without undue complication. In just the same way as the Casualty Officer passes over the more desperately injured victims of an accident to the comprehensive facilities of the hospital, so must the tug pass over a badly perforated, or otherwise complicated case of stranding or grounding to the care of a competent salvage vessel after, of course, providing such interim assistance as may have been necessary to ameliorate the lot of the casualty.

The Commanding officer of the Ocean Tug must be swift to recognise the limitations of his ship and her company in such case, even in the face of a certain commercial loss. He must be prepared to make a rapid and accurate appreciation of any such casualty to which he may be summoned so that, should the accident be beyond the technical scope of his own resources, a salvage vessel can be ordered to the scene without delay. These considerations cannot of course deny the excellent work that has, and certainly will continue to be made by Ocean-Going Tugs in the absence of full salvage facilities.

In the broadest possible sense, it may be assumed that Ocean-Going Tugs are capable of dealing, without salvage vessel assistance, with any vessel which has suffered a simple stranding or grounding with no perforation sufficient to cause a critical loss of buoyancy, and under conditions where, if the time/tide factor is not favourable to an immediate rescue, a suitable draft/trim refloatation condition can be achieved by the discharge of ballast, fuel, stores or cargo, without calling upon outside assistance.

Because this work deals with the function of the Ocean-Going Tug type and the practical aspects of towage, it is not proposed to dwell upon the various operations of salvage required before a casualty may be refloated with or without the assistance of tugs. These processes, operations and manoeuvres are most competently described in the Training Manuals associated with the Navies and Coast Guard Services of the world, in *Lloyd's Calendar*, and variously in the products of the more noteworthy of both British and Foreign publishers specialising in the nautical sphere.

Before proceeding with a description of tug usage in the matter of beach work, attention is drawn to the widely held belief that the standard solution to problems resulting from a stranding or grounding is provided by the application of a sufficient aggregate of tug horse power to literally drag the casualty over, or through, intervening shallows, into deep water. In demonstration of the utter fallacy of this belief, attention is drawn to typical values of bollard pull co-efficients as obtaining in contempory ocean-going types. In the case of tugs fitted with fixed pitch propellers a bollard pull of about ·011 tons for each unit of Indicated Horse Power may be anticipated, whilst tugs fitted with controllable pitch propellers improve upon this performance to values of ·012 to ·014 tons per I.H.P. From this it follows that an average unit from the Ocean Class of Tug would provide, in calm weather and still conditions permitting of a straight and steady pull, an effort in the vicinity of 45 to 50 tons. An average type of foreign-going merchant vessel of about 8,500 gross tons, having taken the bottom whilst fully laden, could offer a problem amounting to the full value of her displacement tonnage, i.e. something in the vicinity of 17,000 tons. If the whole, or the greater part, of such a vessel's bottom is in firm contact with the sea bed it follows that no practically achievable or operable aggregate of water-borne power can move her. Towage assistance in such cases must therefore always be in the nature of towing a partly water-borne vessel away from the shore rather than a true refloatation as such. Tugs cannot therefore be usefully employed in the assistance of stranded or grounded vessels until ship/seabed contact, in the case of the casualty, has been reduced to such an extent, by the reduction of displacement or an appropriate adjustment to the casualty's trim, or by a combination of both with the tidal lift obtaining, as to provide a manageable operation for the total tug horse-power available plus any other assistance such as may be provided by ground tackle and hauling off purchases.

The relative virtues of tugs and hauling off purchases are frequently compared so that a bland comparison is offered here. Remembering the bollard pull coefficients offered above, the pull exerted by a modest windlass equipment with a six-fold purchase resolves as:—

$$W. = \frac{S(8 \times P)}{8 + n}$$

Using, in this familiar formula, a value of 15 tons as representative of a modest windlass capacity we have:—

$$W. = \frac{15\,(96)}{20}$$

W. or power developed = 66 tons.

O

This shows that a quite modest windlass installation, if augmented by a suitable hauling off purchase, can provide a dead pull something in excess of the bollard effort of tugs of some consequence; lest this might be assumed to be denigratory to tug performance, it is prudent to recall that the tug's bollard pull can be exerted continuously and almost indefinitely whilst that deriving from the windlass must be limited to the extent of the hauling off purchase. The windlass cum purchase effort obtains in a series of spasms punctuated by the period required to refleet the gear so that the intelligent salvor relies upon tug effort to maintain the gain obtained in concert with the windlass during the re-fleeting pauses.

Whether as a part of her own salvage processes or as a part of her obligation to provide interim assistance to a casualty whilst awaiting the arrival of a salvage vessel, the Ocean Tug is occasionally required to set out one or more anchors from a casualty. These may be 'Stay' anchors, intended to prevent a vessel from changing her position whilst deprived of her own initiative, or 'Hauling Off' anchors intended to assist with rescue processes. When there is no immediate prospect of refloating a casualty, steps must be taken to ensure that she is not driven into further hazard through tide effect, deterioration in weather or whilst lightening operations are being progressed. It is also very clearly advantageous to all concerned to keep a stranded or grounded vessel as near as may be achieved to a right angles aspect to the shore or to a reef or shoal configuration, both from the point of minimising damage and in the facilitation of an ultimate refloatation. Conversely, if this attitude has been lost by broaching after taking the bottom, arrangements must be made to correct the defect at the earliest possible moment consistent with other requirements.

It is not proposed to dwell upon the seamanlike procedures involved in laying out anchors since this has been done most expertly in works of more merit than this, it only remains to be offered that if the casualty can offer a free side, and if she is lying in a sufficient depth of water, the easiest way to lay off anchors will be with the use of the tug. Even if the tug cannot get right in alongside, her capstan or towing winch power can be readily offered to heave off such floating facilities as may be in use for the support of anchors. The laying out of backing anchors to 'Hauling Off Gear' is likewise very much facilitated if the tug's heavy gear is used.

Before leaving this aspect of beach work the remark must be offered that any complexity of stay and hauling off anchors inevitably provides a dual complication for the tug. The first being that such circumstances may necessitate the tug laying out, and buoying, her towing gear in order to avoid risk of fouling. The second is that the positions of the anchors, and the lie of the attendant gear may seriously restrict the tug's movements during the actual refloatation. Clearly prudent compromise is indicated according to the particular operation.

CHAPTER II

The Utilisation of the Ocean-Going Tug to the Best Advantage whilst Removing a Casualty from a Stranded or Grounded Position.

On every refloatation operation the tug or tugs concerned will be required to exert their towage effort upon an indicated line relative to the casualty's fore and aft line and to her position relative to the shore or shoal and to critical soundings nearby. It is always the first consideration of any tug Commanding Officer that until the ship involved begins to move, or the attempt at refloatation is suspended, he will be obliged to hold maximum tension on his towing gear, on the desired line, so that, notwithstanding the fact that the tug's engines will be turning over at full or near full revolutions, she will in fact be fully stationery and at the mercy of wind, weather, tidal or current streams in very close proximity to the shore. Although the tug's prime function lies in the provision of assistance to others in difficulty, the Tug Commanding Officer's first responsibility is to his own charge so that in rendering assistance to a vessel aground he must pay very careful heed to any hazard involved, making accurate navigational allowances for the physical conditions applying and keeping the most careful check upon his ship's precise position relative to the casualty and the grounding hazard by every means available to him, never hesitating to discontinue operations when the balance of hazard against success goes the wrong way and when this action can be assumed without hazarding other associates in any adventure.

The fact that the casualty is aground implies that all of the associated manoeuvring must take place in relatively shallow water, this necessitates three further considerations, the first is that when the casualty is aground on a soft bottom there will always be a risk of fouling ship's side injection orifices with weed, mud, sand or other sea-bed debris. Another is that any impediment to the free dispersion of screw race detracts from towing efficiency and the last applies if the tug is working close to the casualty. In such case the sea turbulence resulting from a reflection of the screw race back from the casualty's hull, and from the sea-bed, increases propeller cavitation to such effect as to detract very considerably from the tug's performance. It follows therefore that the tug should exert her force as distantly as circumstances will permit from the casualty in the deepest water practically obtainable even when this requires the use of quite extensive scopes of towing gear.

In the consideration of screw effect, but in a somewhat different context, reference must be made to the cutting and scouring of the sea-bed which occurs when large propellers are worked in shallow water with sand, or mud, or shingle bottom. This effect is sometimes put to a good use in providing a dredging function when sand or shingle banks complicate a refloatation but, in the course of refloatation operations under such conditions, the equal and opposite effect must never be overlooked. Inadvertent shoaling often results from working tugs too close to a grounded or stranded vessel on sand or shingle, through the material which is loosened and thrown up against the casualty's hull by the tug's screw race, this can often, in the course of an hour or two of towing, provide a bank of sufficient dimensions to seriously negate the effort expended.

Ordinarily speaking the best towage effect obtains from a straight pull along a defined best line into deep water, but occasionally circumstances indicate the need for working a ship free, usually when she is pinned at one point. The action then required is something akin to that required to extract a cork from a bottle when pressure or tension is applied alternately to one side and the other to pry it loose. This working motion was for many years only appreciated in terms of the lateral movement imparted to the casualty by sheering the tug across from one quarter to another, but more recent investigation indicates a further virtue. The incorporation of a dynamometer into the towing medium of a sheering tug showed that there was a distinct improvement in towing tension engendered at the moment that the tug was swung from one heading to another at the limit of sheer. Whilst there is no fully tenable explanation available in explanation of this condition, it does seem that some leverage advantage may derive from the distance of the point of tow to the turning point; be that as it may, it is a fact that sheering does provide added pull.

Casualties involving large ships, or the more complicated type of refloatation where a great towage effort must be exerted over a limited period, often demand the assistance of several tugs simultaneously. In such case care must be taken to see that tugs do not, when this is practicable, work in one another's screw race. When this is not practicable because of a restricted arc for tug operations then tugs must operate in tandem or, as the older tug men term it, 'Work in Shafts.' When this is obliged it is necessary to provide the second tug in train with towing gear equal to the combined pull. When tugs are required to operate in tandem across any tide of any consequence the combined output is disappointingly low in view of the loss of effort entailed by tide allowance to hold position.

Whenever refloatation towage proceeds using a long scope of gear, and where sheering is involved, the towing medium must be in the effective control of an efficient gog-rope or bridle to make sure that the screw is not fouled at the moment of coming about.

Most operations of this kind are assisted by working the casualty's engines at full speed astern. It is occasionally the case that when the ship comes free the ship's engines are not stopped forthwith so that she gathers sternway. Not only is there a grave risk of fouling the casualty's screw at this moment on either the hauling off or stay anchor gear, but also with the towing gear; there is also real hazard in this for the tug since she may well be overtaken and girded if she happens to be placed broad upon the quarter of the casualty at the moment of release, for a variety of reasons. It is therefore entirely obligatory upon the tug personnel to keep the closest watch upon the casualty so that they shall be aware of her release so as to take the proper steps for the safety of the tug.

In the event that a refloatation is achieved in good order then it is a part of the tug's duty to tow the vessel well clear of danger so that she may swing and set course without further risk. Also, in the event that further towage assistance is entailed, a fair off-shore position is essential before a sea towage rig can be arranged at the forward end.

Refloatation operations impose serious torsional and crushing stress upon the inboard parts of the towing medium, at the casualty end, in the vicinity of bitts and fairlead. In view of this it is good practice to attach a short steel wire pennant of suitable size to the end of the permanent gear to accept this punishment. It would also provide a part of the casualty's contribution to the operation to take steps to reduce this damage so that the Tug Commanding Officer would ensure that the proper and necessary instructions were passed to take care of this.

CHAPTER III

Salvage Equipment Borne in Ocean-Going Tugs.

Having, in Chapter I of this Section, briefly outlined the role of the ocean-going tug in the reflotation of stranded vessels, attention may now perhaps be drawn towards the special equipment which may be provided on board of tugs in order that they may discharge their responsibilities in this direction to best effect. As with all and every commercial adventure, any Operator's inclinations in the matter of equipment will be wholly within the terms of the financial return resulting from the outlay envisaged. It is entirely a fact that an Operator can equip a tug with every device imaginable to improve salvage efficiency, only to find that his vessel is never in any position to make use of them thereafter. It is, however, equally as often the case that a tug cannot take advantage of magnificent salvage opportunities because she lacks certain items of essential equipment. It will, therefore, almost always be the case that salvage equipment in tugs will represent a compromise, tending on one side or the other of a reasonably average equipment, according to the operator's past experience. This chapter endeavours to give a description of the sort of equipment which may well find a place in the modern ocean-going tug.

1. Handling Gear.

Every ocean-going tug is provided with a mast and derrick for handling ropes, stores, salvage gear and the like and the usual capacity for such equipment approximates to a safe working limit of five tons. Power for this derrick is commonly provided from the Automatic Towing Winch, the after Capstan or from the windlass; it is unusual indeed to find a derrick winch installed on board of tugs. Because design considerations necessitate the location of the main mast in tugs at a point well forward of the point of tow, and because other major considerations demand that the forecastle head superstructure shall extend aft as far as the point of tow, any derrick heeled to the main mast must pivot about a point some fourteen or fifteen feet above towing deck level; this whole arrangement presents working conditions which are far from satisfactory, especially bearing in mind the length of derrick necessary to plumb well outboard, and the fact that tugs are occasionally required to transfer gear, stores and the like in a seaway whilst working alongside of a casualty.

It is, therefore, conscientiously offered that the handling requirement, in the salvage function, could be much more adequately served by the provision of a pair of properly proportioned samson posts and derricks located in a position at mid half width outboard of the centre line on each side abreast of, or slightly forward of, the point of tow. Such an arrangement would present a derrick of more manageable length at a reasonable height above the working deck, and much more accessible to deck machinery than any centre-line arrangement.

2. Deck Illumination.

Salvage operations are almost invariably conditioned by tide, rarely however by time, so that operations, once placed in hand, are always progressed by day and by night until a result is achieved. Proper illumination for dark hours working is therefore entirely obligatory. To this end it is customary to provide tugs with a large searchlight, mounted aloft, together with a number of flood lights arranged at strategic points on masts, funnel and upper works. Supplementary illumination is then provided with portable lights in some variety, blocks and halliards being rigged as convenience dictates about the loftier features.

At one time, the humble sheet metal cluster type of light was regarded as being entirely adequate to tug requirements, but recent experience in gas contaminated atmospheres, whilst working upon tanker casualties, has somewhat drastically demonstrated the necessity for gas-proofing all working electrical fittings in tugs.

3. Salvage Pumps.

Most salvage operations include the removal of a greater or lesser quantity of water from one or more compartments of a casualty. A salvage pump, or pumps, represents therefore equipment which is wholly mandatory. Opinion will, of course, sharply divide as to whether one pump is adequate to the function or whether two are necessary. The merits of the reciprocating plunger type of pump will be contrasted with those of the centrifugal patterns and so on but, regardless of number or type, no ocean-going tug's pumping capacity should be less than **2,000** gallons per minute; preferably a great deal more. It is also entirely essential that this capacity should be capable of being directed into more than one compartment simultaneously.

A half a century, or even a lesser period ago, the unhesitating choice for a salvage pump would have been for a large capacity duplex plunger pump installed so as to draw from a single head, port and starboard, and discharging through the bottom. The contemporary choice is, however, for a twinned installation of a pair of centrifugal pumps, diesel driven, of the two or three stage type fitted for suction or supply (Figure No. 56). Such pumps being arranged to draw through one or more swivelling multiple salvage heads suitably mounted or, in the supplementary role, supplying higher pressure water to fire heads and monitors; such flexibility being dependent upon the efficiency of the interior pipe-line sytem.

A pump set such as has been briefly described, would require a suitable equipment of hose and associated components. Standard salvage hose is of 6 ins. diameter (although both smaller and larger sizes are made) in 10 ft. lengths, it is flexibly constructed of canvas and rubber reinforced with spiral, sem-imbedded galvanised steel wire. Each 10 ft. length is fitted with a male and a female metal triple swing bolt coupling to British Standards Institution requirements.

Each pipe-line rigged from the multiple head must be fitted with a priming bend and a foot valve cum strainer. Most salvage pumping operations from tugs entail upward trends of the pipe-lines from the salvage heads to the casualty's rail before they descend to the flooded compartments. Before pumping can commence it is necessary to prime the pipe-line by charging it fully with water from the pump right to the end of the line. This is effected by providing the suction end of the pipe-line with a one way valve called a foot-valve, which is usually associated with a sturdy strainer head. It is also best practice to arrange a relief valve in the foot valve operated by a light line from above so that if it is desired to empty the line, such as prior to unrigging a line, this can be conveniently arranged. To facilitate the charging of the pipe-line with water a priming bend is introduced at the highest point to the line. This bend consists of a 90° galvanised steel 6 in. diameter bend fitted with standard triple swing bolt connections at each end. A standard $2\frac{1}{2}$ in. male connection and an air cock are arranged side by side at the mid point of the outer curve. In service, and before pumping commences, general service water is introduced into the line so as to solidly fill it from foot valve to the pump casing, the air cock being left open to exhaust all of the air. After ensuring that the foot-valve

FIG. 56.

By courtesy of Merryweather & Sons, Ltd.

Inlet: 8″	Outlet: 6″
Gals/min at lbs/sq. in.	
1500 (6820)	100 (7)
1350 (6135)	120 (8·4)
1050 (4770)	150 (10·5)
900 (4090)	170 (12)
2000 (9090)	Salvage duties
NOTE:—Figures in brackets indicate Litres and Kg per sq. cm	

is properly submerged, the air cock is closed and the gate valve to the 2½ in. male connection is closed and the pump is started on a completely filled system.   (Figure No. 57).

Swivelling Multiple Salvage Head on a 5″
Flanged Main.

Salvage Hose Foot-valve
and Strainer fitted with a
Release Valve.

Salvage Hose Priming Bend fitted
with a 2½″ Charging Connection and
Air Cock.

Fig. 57.

Because there are occasionally conditions on board of casualties where the tug's main pump equipment cannot be put to use, it is best practice to provide an equipment of portable pumps together with suitable hose and fittings. There is a considerable range of pumps available so that individual tugs' equipments will vary considerably. The most elementary equipment will consist of a small diesel driven single stage centrifugal pump, operating upon a 4 in. or 5 in. inlet, mounted upon a wheeled bogey fitted for slinging from a derrick. This pump will be placed inboard of the casualty and will be operated as a self contained and independent unit. This equipment does,

FIG. 58.
'Aquator' Portable Suction Device.

however, suffer from a number of operational drawbacks. Even the smallest of these pumps still represents bulk which is not easily manoeuvred below decks so that its application is limited. Such pumps are too bulky to be stowed in tugs' engine-rooms and are, therefore, usually consigned to the Hold or the Salvage Store where maintenance tends to become somewhat irregular to the detriment of the pump's ultimate service performance.

If a smaller output capacity can be accepted, there are other and more convenient means of fulfilling the need for a portable pump facility, one of these is the 'Aquator' suction appliance. (Figure No. 58). This is a simple device which takes delivery of high pressure water from the tug's fire-pump and transfers the energy in this water to a larger volume of water at a lower head. This is achieved through the medium of a high efficiency two-stage ejector nozzle which forms the largest part of the device. The high pressure water from the fire-pump is introduced to the ejector via a standard $2\frac{1}{2}$ in. male coupling mounted upon a base plate, the suction end of the ejector being open to the water at the base and protected by a strainer plate. The swirling energy of the induced water stream generates ejector lift so that both supply water and an amount of salvage water is forced upwards through the 'Aquator' to be discharged through 6 in. salvage hose, secured to the head of the device by means of the standard triple swing bolt attachment, to overside at a delivery rate which will vary according to the pressure available.

This equipment has certain extremely desirable qualities, the first being that it can work constantly and efficiently over long periods because it possesses no moving parts; it is safe to use in small enclosed compartments because it emits no heat or exhaust gas, and because it is only about 2 ft. 4 ins. high and, being made of an aluminium alloy, weighs only 50 lbs., it is easily passed below decks. Perhaps its greatest virtue is, however, that it will operate effectively at depths well in excess of the normal maximum suction lift of about 28 ft.

Another portable pumping device which is in general use is the portable submersible pump which is designed to be lowered wholly within the water to be discharged by means of a derrick fall or purchase. Such pumps are either electrically powered or operate on compressed air so that fair access from the tug is quite essential. Delivery from such pumps is excellent, although, quite obviously, the maintenance of the electrically-powered examples is both critical and difficult.

4.   Self-Contained Diving Apparatus.

The majority of Ocean-Going Tug Operators declined, for a great many years, to equip their vessels with diving gear and to employ the specialised personnel thereby necessitated. There were very good reasons for this attitude, that most commonly quoted being the consideration that the small number of salvage operations falling to tugs which involved diver assistance simply did not justify the acquisition of the extremely expensive gear involved. It was also fairly well known that in cases where Operators had purchased the necessary equipment the paucity of work subsequently obtaining, and the lack of professional practice, led divers to abandon tug employment for more remunerative work elsewhere. Notwithstanding all of this, many Operators were alive to the fact that deficiencies in this direction were subtracting from operational efficiency and earning potential.

Circumstances in this connection began to change with the introduction of 'Skin Diving' following upon the experiences gained in World War II so that diving, in the depths to be associated with the Towage/Stranding/ Grounding function, has become a very popular sport and expertise in shallow water diving is now so readily acquired that a great many of the younger Officers and Ratings are most willing, and able, to serve in the diving capacity, particularly if a modest financial inducement offers. The equipment required, compared to the traditional assemblage of copper helmet, pump, hose and canvas suit etc., is moreover very reasonably acquired whilst operational maintenance is the essence of simplicity.

The equipment in most common use on board of tugs today is the compressed air, self contained, twin cylinder, half-mask, hose and mouthpiece set. This is supplied, in preference to the alternative closed circuit breathing apparatus, because, with slight modifications to the air compressor equipment carried by all motor tugs, to provide the necessary multi-stage effect, and to provide proper filterage to ensure a clean, moisturised and oil-free air supply, an almost unlimited capacity of compressed air for diving operations and practice is made available at minimal cost. This basic equipment may, of course, be augmented according to the wishes of the operator with wet or dry suits, hoods, fins and other accessories.

The normal equipment, as is illustrated in Figure No. 59, has a capacity for 100 cubic feet of free air giving an underwater endurance of approximately 90 minutes upon the sort of duty to be associated with a grounding or stranding operation. As an 'Aqualung' is an open circuit apparatus wherein the exhaled air is exhausted to the atmosphere and not rebreathed it follows that the endurance of any given capacity decreases with increased depth in accordance with Boyle's Law. Endurance will also clearly be proportionate to the diver's physical activities and exertion. The equipment commonly includes a pressure gauge to show the contents of the cylinders but more elaborate equipments provide a device to warn the diver of approaching exhaustion of supplies.

The vital component of any 'Aqualung' is the demand valve mounted usually (and in the case of the illustration) between the air cylinders and directly in line with the upper portions of the diver's lungs. Its function is to control the amount of air to be supplied to the diver according to the demands made by the action of his inspiration conditioned by his exertions.

A single stage demand valve basically consists of a diaphragm, one side of which is exposed to ambient pressure, the other being connected by means

Fig. 59.

Self-contained 'Aqualung' Diving Apparatus.

By courtesy of Siebe Gorman.

of a corrugated hose to the diver's respiratory system. On inhalation the pressure on the diver's side of the diaphragm is depressed and the ambient pressure on the other side causes an inward movement actuating a lever mechanism which, in turn, admits a flow of compressed air into the valve chamber and the inhalation hose. As soon as the diver's demand is satisfied, the pressures on both sides of the diaphragm are re-balanced and the air inlet valve is closed. Exhalation is effected through a second hose and via a non-return valve. It hardly needs to be stated that, in the case of these valves, utmost design excellence and process perfection are essential to ensure that there is no resistance to breathing, especially at peak flows during heavy exertion.

A second type of regulator utilises a two-stage reducing effect. A reduction valve, which may either be spring loaded or open to the sea, is used to reduce the cylinder pressure to a fixed value, usually something between 80/100 lbs. per square inch; air is then supplied to the demand valve at constant pressure. The principal advantage of this two stage refinement lies in that the demand valve can be attached to the face mask directly, thus allowing the air supply to be made through a small diameter high pressure tube rather than the more cumbersome larger bored corrugated hose associated with single stage supply. Paradoxically enough, many experienced divers still prefer the single stage outfit because a supply of air at constant pressure does not, they contend, offer the built-in warning sytsem provided by the single stage set in terms of the operational characteristics resulting from declining pressure.

As with most specialist equipment, there is a great variety of accessories available to augment the performance of the equipment in the way of special warning devices, refinements to demand valves, special reserve air devices and the like, but salvage service, generally speaking, demands a rugged efficiency which is usually associated with functional simplicity.

It is not necessary, in a note of this kind, to discuss the various types of mask except perhaps to observe that whilst the single stage demand valve is commonly associated with the half mask and separate mouthpiece, the two stage equipment will normally require a full mask if the demand valve is incorporated into the mask, to take full advantage of the system.

'Aqualung' cylinders are required to be as light as is consistent with the strength required, this necessitates the employment of a high tensile steel alloy for their manufacture. This material is extremely vulnerable to careless handling. The British Standards Institution recommends that compressed air cylinders should be grey in colour with black and white quartering to the shoulders, *this colour code must, at all times, be maintained clearly and correctly, especially if they are at all likely to be sent ashore for re-filling.* This for the most obvious reasons.

5. Under-Water Cutting Gear.

Largely because the use of under-water cutting gear implied, at one time, the employment of properly trained and expensively equipped and maintained divers, most ocean-going tug operators fought shy of adding this item to their tugs' establishment of equipment. Conservative thinking, in such days, was not moreover wholly impressed with the performance of the relatively narrow selection of gear which was available in any case. The fact that contemporary cutters utilised compressed air shrouding to the cutting flame, which occasioned the employment of three cylinders and three hose-lines with all the nuisance, expense and storage problems thereby involved, was not conducive either to any reduction of sales resistance.

Because of design imperfections, these early under-water equipments were rather wasteful of gases and practically every model, regardless of its country of origin, had tendencies towards back-firing to a destructive, and frequently exceedingly dangerous degree. It was not surprising, therefore, that the general tug view in this matter was that the potential of the gear in use was nothing like equivalent to the costs involved.

This condition began to change when the availability of modern equipment made shallow water diving from tugs a more practicable and economically acceptable proposition, to reach a situation more provocative of changed thinking when the Siebe-Gorman-Kirkham under-water oxy-hydrogen cutter became available.

On the score of economy in initial outlay and in operating and maintenance cost, the 'Vixen' cutter, as it is called, is wholly satisfactory seeing that it is a two hose cutter using hydrogen and oxygen only, without the hitherto requirement of a compressed air shroud. This cutter operates upon approximately 50 per cent of the oxygen consumption of other types in common use and about 60 per cent of their hydrogen consumption. The patented nozzle mixing and the design of the outer shroud is responsible for this economy which detracts nothing from the efficiency of the cutter, or its speed in use whilst it does, in fact, reduce pre-heat time down to two seconds or less.

Although the 'Vixen' cutter is primarily designed for under-water use, it may be used for considerable periods out of water without overheating so that, in the consideration of day to day usage in tugs, there is no need for any duplication of equipment. Figure No. 60 shows the cutter proper which is a

P

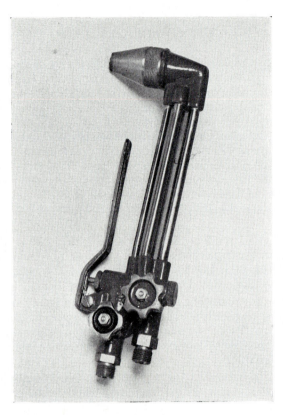

FIG. 60.
'Vixen' Under Water Cutting Tool.
By courtesy of Siebe Gorman.

ruggedly constructed tool which takes down easily for cleaning and maintenance.

One of the most irritating and wasteful defects in the under-water cutting tools available hitherto has presented in the need for surface assistance in re-igniting the gases. In the past this requirement has either led to operational delays amounting to a total of some consequence, or alternatively to a wasteful expenditure of gases if the diver has felt disinclined to extinguish the flame whilst discharging functions auxiliary to the cut proper. The 'Vixen' gear overcomes this embarrassment of alternatives with the provision of a watertight switch and striker plate which leaves full initiative and control with the diver.

6. The Cox Submerged Bolt-Driving and Punching Gun.

This most important under-water working tool was invented as early as 1920 but was not fully developed until 1939, when, for very obvious reasons, it was adopted by the Royal Navy and became a 'secret weapon' to remain on the confidential list until some time after the cessation of hostilities.

The gun, which is massively constructed, principally in gun metal, to a weight of 36 lbs. and to a length of a little under 2 ft. 6 ins., is arranged to break down into two major components, viz. the holder and the bolt, firing block and barrel assembly, for safety and convenience when in actual use. The function of the gun being to fire screwed bolts into steel plates leaving sufficient end protruding for the subsequent affixation of patching materials or other salvage implements. Alternatively the gun may be loaded with a blunt missile to punch holes in plating for a variety of purposes. A third use for the gun is to enable an air bolt to be passed through the shell plating of a sunken vessel in order to facilitate the subsequent admission of air into it for the purposes of reflotation.

All of the bolts fired by the Cox gun, the bolt sections of the missiles and the air bolts, are of a uniform diameter of $\frac{5}{8}$ in.; they are $4\frac{1}{2}$ ins. long and are constructed from heat treated alloy steel. Depending upon the thickness of steel to be penetrated these bolts are driven by dense, fast burning explosive charges suitably graduated and contained in metal cups which function as driving pistons in the gun after ignition. For instance the charge required to drive a bolt into $\frac{1}{4}$ inch thick mild steel plate is graded No. 2 and consists of 22 grains of explosive. Plating $\frac{3}{8}$ in. thick requires a No. 3 charge which is proportionately heavier and so on.

In practice the Diver goes below with the holder component only and the bolt, firing block and ammunition are loaded into the barrel on the surface. With the diver in position and all ready to fix his first bolt, or punch the hole, or insert the air bolt, the barrel assembly is sent down to him whereupon he fits it into the holder where it is retained, against a buffer spring, by means of a catch. When he is ready to fire, the diver places the nose of the barrel squarely against the plate to be punctured, presses the safety catch and then thrusts the holder smartly forward so that the firing pin strikes the detonator to ignite the charge. Momentarily the screwed bolt or missile is held to the firing block by means of a steel tension member but this snaps when the propellant has developed its maximum power and the piston containing the charge shoots the bolt forward. By this means the bolt achieves the necessarily high velocity for effective penetration in a very short distance of travel.

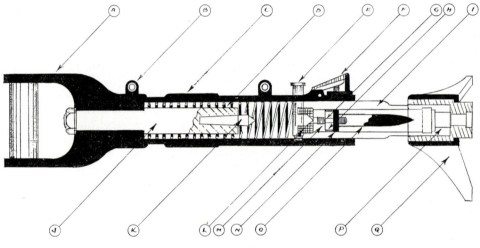

Fig. 61.

Cox Submerged Bolt Driving and Punching Gun.

(a) Gun-metal Holder. (b) Slinging Lug. (c) Re-inforcement Band. (d) Buffer-Spring. (e) Firing Catch. (f) Barrel Retaining Catch. (g) Tension Member. (h) Propellant. (i) Barrel. (j) Buffer. (k) Firing Pin. (l) Breech Block. (m) Firing Block. (n) Piston. (o) Bolt. (p) Piston Arresting Block. (q) Stabiliser.

Simplified sketch by author from material provided by Siebe Gorman.

The bolt comes to rest with the curved part of its nose inside of the plate with the screwed part outside. The shank of the bolt is so firmly gripped in the plate that practical tests reveal that a pull of up to ten tons is required to extract it. Where it is necessary to attach bulky patching material or foundation fabric of heavy section to plating, extension bolts of 1 inch diameter and varying length are available for the purpose.

When it is necessary to supply air to an otherwise inaccessible compartment in order to force out water, the procedure outlined above is carried out with a special air bolt. This bolt is hollow and is fitted with a detachable screwed, solid nose. After the bolt has been fired a special adaptor is screwed to the outer part and, by passing a screwdriver through the centre of both the adapter and the bolt, the nose can be unscrewed so that it falls inward into the compartment and a compressed air supply hose can be attached. The whole operation can be completed without admitting any appreciable amount of water into the compartment and wholly without access to the inner side of the plate. It is important to observe however that the orifice obtained by the gun is not normally adequate to the demands of breathing air. (Figure No. 61).

7. Power Supply to Casualties.

Tugs are very often in demand for the purpose of supplying power to vessels which are not self-sustaining for any one of a great variety of reasons. This facility is provided in a variety of ways according to the requirements of the Operator and the skill of the Designer.

In its most elementary form, electrical supply from a tug may be arranged directly from her switchboard to a suitable connection or distribution point on board of the casualty according to the circumstances obtaining, via a pair of flexible leads run out through the engine-room skylight. Clearly any primitive arrangement of this kind must be provocative of danger and accident prone so that best practice demands arrangements of a more competent order.

Demands of this nature occur so often that many Operators call for an auxiliary supply switchboard to be mounted at a convenient point on the main or upper deck, the Towing Winch House providing a favourite location. Such a switchboard is often arranged so that supplies may be directed towards it from each generating facility borne in the tug, the necessary fusing arrangements and instrumentation being fitted bearing exposure in mind. Supplies from this point are carried via a run, or runs, of armoured and insulated cable properly rigged and slung in insulated material. The flexibility and capacity of diesel-electrically powered tugs, in this direction, is well established. There are at least two occasions where such tugs have provided electricity for sizeable communities and undertakings under circumstances of crisis.

The older generation of steam powered tugs was somewhat more flexible in the supply role than are their diesel engined successors because a supply of live steam at boiler, or any lesser, pressure could be readily provided via the proper boiler-room facilities and flexible steel hosing. With such facilities, power could be supplied to work individual items of deck machinery, or indeed a vessel's whole equipment if necessary, a capacity which cannot possibly be emulated by any compressor equipment likely to be mounted in a tug.

Another facet of the supply function often required of tugs is that of water tender. A tug's utility in this direction must obviously be qualified by the excellence, or otherwise of her pipeline plan, but it is apparent that with the tankage available for this service and the equipment of pumps normally encountered, most tugs should be capable of rendering good service in this role.

8.   Fire-Fighting Equipment.

Maritime Salvage history is liberally be-sprinkled with instances, both in war and in peace, where ocean-going tugs have rendered yeoman service in the fighting of shipboard fires in port, in anchorages and roadsteads, and upon the high seas. The ability to deal competently with the fire hazard is now a traditional part of the sea rescue function and a comprehensive outfit of fire-fighting appliances has become an essential feature of the ocean-going tug's equipment.

Efficiency in maritime fire-fighting invariably stems from the ability to project an ample and flexible quantity of water towards the seat of conflagration the while retaining a sufficiency of service to protect the tug herself from the hazards attendant upon close assistance. In the modern ocean tug this demand is fully satisfied by the dual service which is returned by the centrifugal salvage cum fire pumps which are now almost universally fitted. This type of pump with its two or three stage design, coupled with other design excellences, is able to provide a flexibility of performance which will produce the high pressures necessitated by foam generating systems for fighting oil fires and the like, the vast quantities of water at medium pressures for the more straight forward aspects of dousing fire and for preventing the spreading of fire, for cooling and protective functions, besides handling the very large quantities of water at relatively low pressures required by salvage assistance.

In addition to this basic equipment some operators provide a separate fire pump which may either be installed in this role independently and exclusively, or which may consist of one component in a combination pumping/generating plant. These will, of course, be supported by the tug's general service pump and deck service pipe lines. Regardless however of the detail of the equipment, the principal consideration in the matter of pumping facilities lies in the provision of a pipe-line system which will allow of the most flexible application of the power available.

FIG. 62.

Quadruple Delivery Head on 6″ Diameter Flanged Inlet. Swivelling 2½″ Standard Outlets on individual Gate Valves.

By courtesy of Merryweather & Sons, Ltd.

The actual equipment of fire-fighting gear will comprise a selection of fixed and portable appliances represented by monitors and hoses and their accessories to supply either water or foam. Most modern tugs mount at least four monitors and heads for eight to twelve hoses, *and this standard of equipment must not, by any means, be regarded as erring upon the generous side, rather is the reverse the case.* Recent experience with a tanker fire showed that

Main Supply Detail for
Multiple Fire-Fighting Delivery Heads.

Number of Outlets	Diameter of Flanged Inlet in Inches
2 Way	5 ins.
3 Way	5 ins.
4 Way	6 ins.
5 Way	6 ins.

Fig. 63.

the full use of a tug's equipment of twelve hoses and four monitors, besides the tanker's own equipment, could only barely hold the conflagration in check pending the arrival of other vessels The scale of equipment will depend of course upon the capacity of the pump establishment but Figures No. 62 and 63 indicate that four standard $2\frac{1}{2}$ inch hoses require a 6 inch supply, and Figure 64 shows the necessity for a 4 inch supply for each monitor so that the problems associated with supply and demand are fully resolved.

It is a considerable advantage in tugs to locate fixed fire appliances as high in the superstructure as may be contrived. In the case of monitors, it is becoming increasingly the practice to mount these upon mast platforms and where tugs mount a pair of samson posts in lieu of a main mast these are often designed to provide support and accommodation for a most effective multiple installation.

The main considerations in the location of deck delivery heads is that they shall be clear of other functional equipment and that there should be an adequacy of manoeuvring space about them for the handling and proper disposition of hoses. Favourite locations for deck delivery heads are on the upper deck both forward and abaft of the bridge superstructure. Wherever space is restricted swivelling heads are quite essential and the provision of a gate valve to each outlet is necessary if full freedom of action is to obtain.

Monitors are manufactured in 4 in., 3 in. and $2\frac{1}{2}$ in. sizes, the figures relating to the diameter of the supply line. All of these may be used for either foam or water but a dual purpose monitor, capable of changing

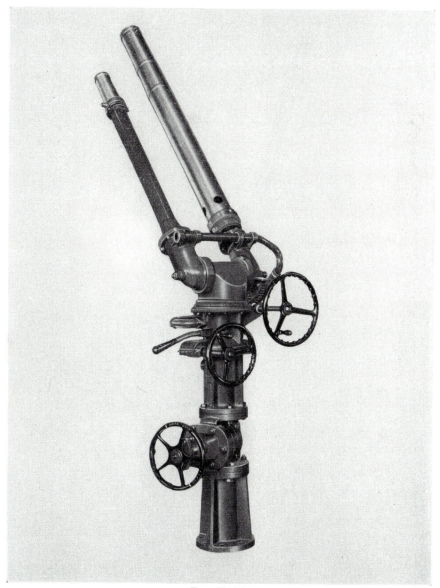

FIG. 64 (a).
4″ Dual-purpose Monitor.

By courtesy of Merryweather & Sons, Ltd.

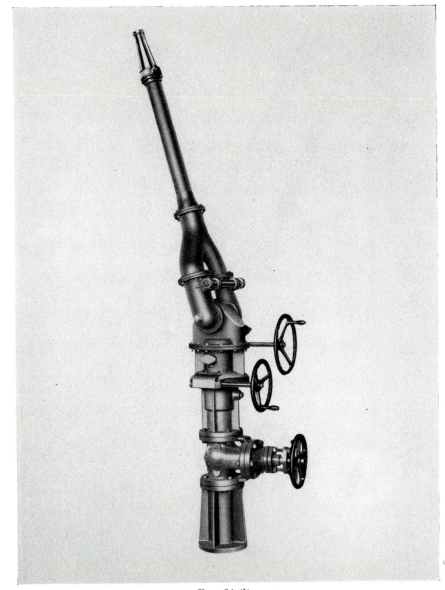

Fig. 64 (b).
4″ Foam or Water Monitor.
By courtesy of Merryweather & Sons, Ltd.

Maximum Capacity Handled			Gallons (*litres*) per minute
Size	Water 100 lb./sq. in. (*7 kg/scm*)	Water 150 lb./sq. in. (*10·5 kg/scm*)	Foam
Dual purpose	1750 (7955)	3150 (9774)	3000 (13638)
4″	1750 (7955)	2150 (9774)	1000/2000 (4546/9092)
3″	1000 (4546)	1300 (5910)	1000/2000 (4546/9092)
2½″	700 (3182)	850 (3864)	400/1000 (1818/4546)

instantaneously from foam to water by the movement of a change-over lever, is manufactured in the largest sizes. Foam-making branch pipes are available for use with the hoses.

There is, of course a variety of hose nozzles available to meet the various and varying requirements of fire-fighting and a selection of these would provide an important part of every tug's equipment. 2½ inch canvas fire-fighting hoses, with British Standard instantaneous couplings, are the standard equipment of all British tugs so as to facilitate the fullest co-operation with shore fire-fighting services should the need arise.

In any discussion of shipboard fire fighting, and the appliances involved, it must always be borne in mind that this aspect of sea rescue is probably the one which places the tug, and her personnel, in the most protracted, if not the most desperate, hazard. It follows, therefore, that some part of water delivery potential must be reserved for the protection of the tug herself. This is commonly achieved through the use of the general service pump and the deck service pipe-line in combination with special patented appliances designed to either produce a flat curtain of water between the tug and the object of her attentions, or to project a heavy cooling spray for the relief of overheated surfaces on the tug or areas of the striken vessel adjacent thereto. Figure Nos. 65 and 66.

Fire in ships, particularly in tankers, is often productive of dense clouds of smoke which not only impedes fire-fighting operations but also tends to produce a separate class of casualty. A well-found tug should, therefore, not only carry an adequacy of resuscitation devices to succour such casualties, but should also be provided with respiratory aids in such numbers as are commensurate with the number of appliances which can be brought into use

FIG. 65.

Device to produce a heavy Water Spray.

By courtesy of Merryweather & Sons, Ltd.

FIG. 66.

Device to produce a flat curtain of water.

By courtesy of Merryweather & Sons, Ltd.

simultaneously. A modest provision in the way of flame proof clothing would also clearly augment fire-fighting efficiency.

The vast number of tankers in service and the constantly increasing size of individual units in such service, together with the profusion in variety of oil derivatives and other liquids having a low flash point which are now being transported in great bulk at sea today, tend to produce something of an increased fire hazard at sea, particularly in narrow seas and coastal waters where traffic tends to congest. Practical experience would appear to insist that this trend should be reflected in all new ocean tug construction, by the installation of proper tankage for at least one hour's run on foam-making compound if fire-fighting resources are to measure up to the sort of problems which must be envisaged in connection with this degree of development.

SECTION VI.

CHAPTER I.

Contract Towage.

Generally speaking, the modern ocean-going tug will find as much employment with contract towage, by which is meant the transportation by towage of ships or other inanimate floating objects in some variety, as she will with salvage or rescue operations. Indeed, it is generally accepted within the calling that it is only the income which derives from these sources which permits the long term maintenance of other tugs, upon distant stations, in anticipation of the more lucrative awards which are associated with marine casualties.

The conditions governing the employment of tugs upon contract operations will clearly vary a very great deal from operation to operation, and according to the customs and practices of the parties involved, but ordinarily the responsibility for preparing any vessel or object to be towed rests with the Owner of the vessel or object. Besides the provision of a tug, or tugs, competent to the operation, in terms of both power and endurance, the tug operator is commonly obliged to provide towing media in such numbers and of such standards as will satisfy the other contracting party. When the nature of the object to be towed indicates the necessity for the use of a towing bridle, the responsibility for supplying this item of equipment may be the responsibility of either party according to agreement. In any case, it is a common condition of contract that all equipment, preparations and provisions must be subject to survey by both parties to the contract.

It is therefore apparent that successful contract towage operations depend upon the conscientious fulfilment of the following three broad terms:—

1. Full seaworthiness of the vessel or object to be towed.

2. Towing media of full competence to the occasion.

3. Tugs of power and endurance entirely adequate to the operation.

226

1. Seaworthiness of the Vessel or Object to be Towed.

It is wholly incumbent upon the owner of the vessel or object to be towed to provide documentary evidence of satisfactory seaworthiness for his property from a source acceptable to the Tug Operator, this is particularly important in the case of a vessel requiring towage following upon an incapacity resulting from accident.

When vessels or objects for towage are relatively small or have little freeboard, so that they are more than ordinarily vulnerable to sea damage, it is necessary to rig substantial temporary weather protection in the way of extensions to bulwarking, closures to hull and superstructure openings, etc. Floating objects, such as cranes and dredgers, which have functional features and characteristics such as extensive hull openings and appendages, or top weights at considerable height and of some consequence, must be properly compensated down to the point of partial dismantlement, if seaworthiness and stability are likely to be impaired. A prudent Tug Commanding Officer would see that he was associated in such considerations at the earliest practicable time. The provision of an entirely satisfactory point, or points, for towing would, in all such operations, figure prominently in the preparations to be made.

Practical experience at sea in the contract towage of normally shaped vessels would appear to indicate that no advantage derives from the utilisation of both of a ship's anchor cables in the form of a towing bridle. There is however evidence and to spare in support of the use of a single cable for this purpose because of the flexibility and efficiency which such usage allows; whilst the fact that this practice leaves one anchor and a full scope of cable readily available in a hawse-pipe has moreover much to commend it. Evidence also exists to permit the recommendation that any anchor displaced for the purpose of making a cable end clear for towing can, to great advantage, be rigged up to an Insurance wire to serve in an emergency.

Whilst discussing the utilisation of anchor cables for towage it may be remarked that it is considered good practice, upon contract operations, to provide a second emergency connection on board of the vessel to be towed. This is most conveniently achieved by detaching the first two shackles from the cable selected for towing and to turn this up to at least two pairs of bitts at one end, allowing sufficient scope to the other as to allow it to reach down to within a fathom of the water-line. This end is passed out through one of the forward leads and brought back inboard over the guard railing or bulwarks, to be provided with a ready greased and eased shackle. The bight of this chain is then stowed conveniently on the forecastle-head deck, stopped up clear of working areas.

The outboard end of the third shackle is then rove over the cable compressor and through the hawse-pipe to be lowered to a position appropriate to the working level on the tug.

If the circumstances of a projected operation indicate that the stress of towage is likely to be great, such as with the towage of a large vessel having a foul bottom, or other comparable condition, it may be considered advisable to rig some form of relieving gear to the cable to be used. From the purely practical view-point, such gear is of little use unless it can be set up taut following upon any adjustment to cable length and as this condition practically predicates the provision of auxiliary power, it follows that its application is somewhat limited. Any gear so rigged must clearly be of strength equal to the towing medium proper and usually takes the form of a steel wire rope purchase shackled into the cable immediately forward of the cable compressor, the hauling part being led off and made up to bitts. When a cable compressor or a windlass is suspect or damaged, this type of gear may be set up to some advantage to the cable on the after side of the cable holder, but in such case the lead of the tackle must be most carefully considered so as not to reduce whatever efficiency remains in the windlass.

In all of this kind of preparation, contract operations possess an immeasurable advantage over salvage or rescue operations in that the necessarily arduous work can be performed in port under calm and still conditions, with every facility to hand and with a sufficiency of labour always available.

As has been offered elsewhere, considerable advantage derives from disconnecting the tail end shaft thus allowing the propeller to revolve as the tow gathers way, so reducing resistance to forward movement. Obviously, the disconnection must be effected forward of the thrust block. Also, if no auxiliary power is available, the rudder should be centred, making sure that the hand-gear is in working order so that helm alterations may be made if circumstances indicate such a need.

Floating objects of other than ship shape tend to oscillate about a single point of tow whilst under way, particularly if their frontal areas are proportionately large, so that it becomes necessary, in practically every case, to tow such objects using a bridle if this embarrassing and highly destructive tendency is to be contained.

Floating objects which include towage as a part of their function are ordinarily well and properly fitted for the purpose but it is occasionally the case that other objects, anticipating towage upon one occasion only, may not be so well provided. It is therefore quite essential to see that towing points on board of all inanimate objects requiring towage are carefully and effectively worked into main members of construction in full cognisance of the stresses to be endured and making full allowance for the dimensions of the towing medium components which will be offered up to them.

Towing points must be arranged so as to be fully accessible at all times throughout a towing operation, bearing in mind the weight and general unhandiness of the gear involved, and so that the bridle legs and shackles will not bear unnecessarily or unfairly upon other construction detail. When this latter consideration cannot be satisfied, fairleads of substantial weight and appropriate shape must be arranged precisely according to the detail of the bridles, etc. to be used.

In view of the weight of towing media and bridles, and particularly if any towage operation includes any number of staging interruptions, it is of considerable practical advantage to rig a tripping purchase to the junction of the bridle legs from a position at some height above the towing points. In many objects requiring towage this may be conveniently arranged from superstructure features, but when this is not the case, the resulting convenience and added efficiency most adequately justifies the installation of a substantial davit, or light derrick and post, for the purpose.

If it is at all practicable to provide the proper accommodation and services, all vessels and inanimate floating objects under towage must be manned effectively, and in considering the numbers to be embarked, this should always be related to the number required to effect a re-connection in the open seas with the facilities available. If this condition is satisfied then the requirements of the International Regulations for Preventing Collisions at Sea in the matter of navigation lights, day signals and fog signals, for Vessels being towed, and the ordinary practice of seamen in the matter of look-out and communication, can be satisfied without strain. It is also desirable, though clearly not often possible, for vessels and floating objects in such case to be provided with auxiliary power for working steering gear and windlass and an emergency pump.

It is a great practical advantage to be able to ascertain, at all times, the precise distance astern of a tow from the tug, this for the purpose of ascertaining the depth of sag in towing media besides the more obvious purposes. This proceeding is greatly facilitated if large patches of distinctively coloured paint can be applied to positions on the hull and superstructure elements at a precise distance apart, the distance chosen being convenient to the tangent

Q

table. The utility of this practice is augmented considerably if arrangements can be made to show lights at the chosen locations for after-dark use.

2. Towing Media.

All tugs of recent construction are equipped with automatic towing winches which may be of single or duplex construction. It is obvious that, if his tug possesses this valuable piece of equipment, any Tug Commanding Officer will insist upon its use because of the utility, convenience and efficiency which results. Seeing that these winches are designed to accept wire ropes of a specific diameter and length, the gear to be used, upon each and every towage operation, will be that which is stowed upon the winch barrel. This clearly implies that, upon occasion, gear well in excess of actual requirements will be in use, but this condition is always accepted in view of the balance of advantage which accrues. Should the winch gear be too heavy for the actual connection to a small tow then a lighter pennant or other reducing component is introduced.

When no winch is fitted, or when two or three objects are to be towed when only a single barrelled winch is available, then recourse must be made to combination towing sets made up of steel wire ropes together with manila or nylon springs. In such case, and only in such case, would the provision of special towing sets for a specific towing operation be contemplated. If these were not available from the tug's normal equipment as suggested in Section III, Chapter IV, then assistance and guidance might well be obtained from a reference to the Admiralty Manual of Seamanship, Volume III, pp. 130/138 and Figures 6—10 to 6—14. If however the conditions and examples enumerated therein are inappropriate to the intended operation, then the Tug Commanding Officer will be obliged to make up towing sets according to his own experience or to such aggregate of accrued experience as may be available from his Operator's records.

This provides probably the most convenient occasion for offering the recommendation that every Tug Commanding Officer, and every Tug Operator, should meticulously record the circumstances of all aspects of every towage operation for future reference.

When towing bridles form a part of media to be used they are made up in dimensions appropriate to the occasion from studded link anchor cable or from flexible steel wire. They are always made up of separate components shackled to a common ring and *never with* the material made up on the bight.

The larger floating objects are commonly towed upon a 'Y' bridle comprising three equal lengths of chain shackled to a common ring. Two legs are used to effect the connection to the object towed using suitable pin and pellet shackles whilst the third leg adds desirable weight and length to the towing medium proper. Such a bridle is intended to be turned about at staging points so as to ensure that no one component is supporting an excess of strain over the other such as may induce metal fatigue.

In the consideration of strength in towing bridles it must be fully appreciated that the twin legs of a bridle do not always share the burden of towage equally. When the object in tow sheers, or when substantial alterations to course are effected, or during periods of hard wind from a constant direction, it is often the case that the whole weight of towing is borne by one bridle leg only, sometimes for considerable periods. It is therefore quite essential that every component of a towing bridle should be equivalent to that of the main towing medium chosen. This proviso clearly includes the common ring, the shackles and thimbles used and indeed the points of tow themselves.

A prudent Tug Commanding Officer, contemplating any contract towage operation, would see that all solid components associated with his chosen towing media; cable, cable joining shackles and ring, were freshly annealed, this in view of the known effects, in the matter of fatigue to such items, of towage stresses.

The length and proportions of towing bridles are of considerable significance to the success of towage operations; because each leg of the bridle has to be of a strength equivalent to that of the principal component there is no advantage to be derived from reducing the included angle to any particular value in the search for an economic diminution of materials, rather is the reverse the case seeing that, to exert its best corrective effect in reducing yaw in the towed object, a bridle requires a fair span. Ordinary mechanical considerations impose a well-defined upper limit to the included angle, but average service experience would appear to favour an included angle of 60° to 80° between bridle legs for most objects requiring towage.

3. Tug or tugs of a power and endurance entirely adequate to the operation.

(a) Power.

It will doubtless come as a surprise, to many who are newly interesting themselves in sea towage in its various aspects, that the principal guidance in estimations of towage performance in the case of ship-shaped vessels derives almost exclusively from the trials and experiments conducted by William Froude almost a century ago. It will also probably not pass without remark that there is, for all practical purposes, no real guidance available whatsoever concerning towage performances in the case of other than ship-shaped floating objects other, that is, than the trial and error information which reposes in the confidential records of the chief exponents of towage expertise and which, very naturally, provides the basis for that expertise.

In passing it may be fairly confidently stated that whilst expert estimations of power for ship towage operations usually tends towards the provision of power somewhat in excess of actual requirements it is rare indeed for estimations to exceed requirements in the case of other than ship-shaped objects under towage.

In the case of the towage of whole and undamaged ships at sea and in terms of fair weather and reasonable trim, William Froude offered that the resistance offered by water to the forward motion of ships of orthodox form was principally supplied by skin or surface friction augmented by certain elements of eddy resistance and wave-making resistance. Froude considered that, in ships of proper form, eddy resistance was so slight as to be practically imperceptible and that wave-making resistance, at the speeds associated with the vast proportion of ocean towage operations, was practically negligible. He offered that surface friction effect, conditioned by the speed of advance and the condition of the surface, was therefore the principal factor in the matter within the terms of the formula:—

$$\text{Resistance in lbs.} = fSV^n$$

Where:—

S	=	The Area of the Wetted Surface in square feet.
V	=	The speed in knots.
f	=	A Coefficient.
n	=	An Index.

The value of the component S is readily obtained, given the necessary detail of the vessel to be towed, by the application of one of the well known wetted surface formulae such as:—

$$S = 1{\cdot}7L \times d + LB \triangle c.$$

Where:—

L   =   Length of the ship in feet.

d   =   Draught (mean) of the ship in feet.

B   =   Beam of the ship in feet.

$\triangle$c   =   Coefficient of fineness.

Appropriate values for the coefficient and index respectively are however clearly areas for concern. Froude's experiments, reinforced by a deal of practical sea experience, would seem to indicate that, at the speeds normally available for towage operations, the index for use would be 2, or the square of the speed; but where speeds of advance in excess of average, such as upwards of eight knots are anticipated then some increase of index is necessary. The most eminent authorities have utilised f values in the vicinity of 0·1, but this would necessarily require some upward adjustment in the case of vessels which were fouled to any extent. Practical experience in the towage of vessels not too recently drydocked suggests that, in such cases, the f value mentioned might well be increased by so much as fifty per cent.

To the product of the formula must be added the force required to overcome the resistance offered by the propeller, if this cannot be allowed to freely revolve by uncoupling or by total removal, also the resistance provided by submerged portions of the towing medium.

Example.—Required tow-rope pull to tow a vessel of 8,500 tons gross, 5,000 tons nett., 17,000 tons Displacement. Length 475 ft., Beam 60 ft., Mean Draught 27 ft. 6 in. Block Coefficient ·75. Propeller Diameter 17 ft. 6 in. Vessel with clean bottom, trimmed to an even keel. Required fair weather speed in tow 5 knots.

$$\text{\textit{Resistance in lbs.}} \quad = \quad fSV^n$$

$$1 \cdot 7 \ Ld + LB \ \triangle c. \quad = \quad 43{,}580 \text{ square feet.}$$

$$\text{Resistance} \quad = \quad \cdot 01 \times 43{,}580 \times 5 \times 5 \text{ tons.}$$

$$\frac{}{2240}$$

$$= \quad 4 \cdot 86 \text{ tons.}$$

Propeller Resistance.

$$\text{Propeller Diameter}^2 \times \text{Speed}^2 \times 1$$

$$= \quad 17 \cdot 5 \times 17 \cdot 5 \times 5 \times 5$$

$$= \quad 7650 \text{ lbs.} = 3 \cdot 4 \text{ tons.}$$

Towing Medium Resistance.

45 fathoms of $2\frac{1}{2}$ in. chain cable, 200 fathoms of steel wire rope and 60 fathoms of 18 in. manila cable laid towing spring.

Estimated 10% of Ship Resistance . . . Say ·5 tons.

Total Pull Required for Five knots in Fair Weather=8·76 *tons.*

The responsibility for the towage of this vessel would ordinarily be entrusted to one of the Sea-Going Class of about 2,000 to 2,500 Brake Horse Power.

(b) Endurance.

It is a fact that, with modern diesel powered tugs operating at the restricted engine revolutions which the majority of towing operations oblige, most Tug Operators are more usually concerned with endurance in terms of the more domestic aspects rather than in terms of bunkers. To this end most contract operations are planned in terms of the consumption of fresh water and fresh provisions allowing for a handsome margin for each stage of an operation, particularly those where a history of bad weather may have adverse effects upon anticipated performances.

Whilst towage operations have been planned in which tugs have replenished stocks of fresh provisions, fresh water and bunkers from the vessel or object being towed, this is not always practicable nor indeed is it desirable. Present tendencies with tows in the longer categories tend toward staging even when this is not strictly wholly necessary, if only for the purpose of resting and refreshing the personnel involved.

In the matching of any intended towage operation to the performances of the tugs available, due account will always be taken of the weather conditions to be encountered on passage, but besides this further account must be taken of the staging detail. In the case of the example quoted earlier it would, for instance, be wholly practical to allot such a task to a single tug for a straight-forward port to port ocean tow, whereas if the same vessel was required to be towed to a very distant place involving perhaps a canal transit besides other staging calls, the task might well be much more economically performed by two smaller units which would cope with the close quarters manoeuvring required without outside assistance.

CHAPTER II.

Towing Two or Three Units.

Whilst the contract towage of single units by single tugs should not ordinarily present any difficulty or complication which has not been covered in other parts of this work, it is quite apparent that the more complex operations involving either two or more tugs, or two or more units in tow, simultaneously, does merit some further discussion.

1. One Ocean-Going Tug Employed in the Towage of Two or More Ships or other Floating Objects.

It is a relatively common contract circumstance for a single tug to be called upon to transport two units by towage simultaneously, and instances are upon record of such obligations being discharged, to the satisfaction of all concerned, in the case of quite large units over very considerable distances. Demands for the simultaneous towage of three units also occasionally arise, but practical experience regarding exercises more complicated than these does not obtain in sufficient number as to allow of either comment or recommendation in this work.

In the consideration of all of these multiple towage exercises the most important factor, after the selection of the tug to be used, will be the disposition of the units to be towed in terms of the number and arrangement of the towing media to be employed. If it is to be assumed that each unit to be towed will be streamed upon a separate medium then it follows that a tug fitted with a duplex winch will be able to accept a tow to each winch barrel with a third, if so required, being towed from either a towing hook or a bollard according to her equipment. A tug fitted with a single barrelled winch would use this facility for one tow and could tow another unit, or if necessary another two units, from either a bollard or a pair of towing hooks. A tug which has no towing winch however would find the towage of three units simultaneously a very difficult proceeding unless she was provided with a cruciform bollard of suitable dimensions.

Circumstances do occasionally arise which oblige the towage of two units on one towing set, the practice being known as 'In Line' towage. This consists of passing the tug's tow-rope proper to the first of a pair to be taken in tow and then rigging a medium between this unit and another astern, the connecting link being arranged to a size and length appropriate to the unit concerned and the weather anticipated. There are however certain limitations to this method which are deserving of mention.

(a)   The towage of more than one unit upon a single medium engenders an overall length which causes embarrassment whilst navigating in congested waters where, besides traffic considerations, complications often arise as a result of tide, current or weather allowances.

(b)   The shock loading of the towing media is complicated by the seaway movements of three units, especially the inhibited movement of the first vessel or object in train, so that the risk of tow-rope failure is heightened somewhat in comparison to single towage.

(c)   Tow-rope chafe about the after parts of the first vessel in train is difficult to contain so that the implied hazard is a constant preoccupation throughout the voyage.

(d)   Towage efficiency tends to suffer throughout the operation due to the extreme difficulties attendant upon effecting adjustments to the length of the inter-tow connection.

(e)   Both the streaming and shortening in of 'In Line' towing complexes become involved proceedings when they take place in tidal or congested waters. The operation of 'shortening in', in particular, becomes difficult because there is no way of preventing the inter-tow connection falling to the sea bed when the tow is brought to a halt, so that there is almost always the risk of fouling the sea bed with all of the delay and complication which results.

For these, and other related reasons, the calling tends to avoid 'In Line' towage operations as much as is possible. Clearly however circumstances must arise which enjoin this method, either because of inadequacies in tug equipment or because of specific contractual requirements. But, when any initiative in the matter resides with the tug, it is the general rule that 'In Line' towage is restricted to operations of modest endurance in good weather through the lesser populated waters, preferably where ports have deep and uncomplicated sea approaches.

When, as is the more general rule, towage operations proceed utilising a separate tow-rope to each unit towed, the most obvious requirement is to ensure that collisions do not result between the units involved. This is, in the first instance, satisfied by streaming each unit to a different scope of gear, but this simple expedient cannot, of itself, provide a whole solution seeing that it is also necessary to veer a sufficient scope of medium to the nearest unit to the tug as will be appropriate to the weather and other conditions affecting the operation. It is also of classic importance to make sure that the sag in the media to the more afterly units in tow is sufficient to allow the more forwardly elements to pass freely over them. When all of the vessels, or objects, comprising the train are identical, then because they will proceed at the same speeds in tow, their resistances to forward movement will be identical so that the catenaries in their respective towing media will be directly proportionate to their length and any unit towing to a shorter scope will pass safely over another towing from a longer one. When dissimilar objects form the burden of a tow however, the problem becomes somewhat more complex.

In this connection it is self-evident that vessels or objects of such shape as to offer a high degree of resistance to forward movement will induce a smaller catenary in any given scope of towing medium than will a finer lined vessel or object towed at the same speed. Some part of this effect may be estimated or calculated . . . (If the equipment and circumstances allow) . . . as was described in Section III, Chapter VII, but in all cases the objects or vessels which provide the deepest catenaries must necessarily occupy the afterly positions in the train. Here again it is apparent that accurate records of past operations must provide the most valuable assistance in organising these multiple towage operations.

In actual practice in these operations, the usual drill is to form up a towing train based upon the premises above described, then when the train has put to sea, the situation is reviewed in the light of the positions which the units of the tow assume under towage conditions. Distances, catenaries, and the like are then checked by measurement, either by radar, rangefinder or by vertical angles, depending upon the facilities to hand, and the necessary adjustments effected in the light thereof. Whilst this is in process it is customary to make a close inspection of the towing media under stress, which may assume positions and attitudes somewhat different to those which were initially allowed. It is quite essential at this stage to see that some allowance is made for relative changes in position in the train which may be caused, in vessels and objects having a high L/B proportion, by pitching and yawing. (Fig. No. 67).

When the units in tow utilise towing bridles it is not practicable to make adjustments to induce permanent sheer in order to augment any of the effect aimed at by the means above described, although it is a fact that dissimilar objects in tow, whether using bridles or not, do tend to assume positions out of line, the one with the other. When the units comprising a tow are all ship-shaped vessels it is possible to derive some sheer benefit by using alternate cables, port and starboard, in progression from forward to aft, a condition which may be assisted by the application of very small angles of appropriate helm. Needless to say, most of the complications which present in multiple towage operations may be reduced, if not wholly removed, if power is available to allow of adjustment to the towed end of the towing media, or for steering.

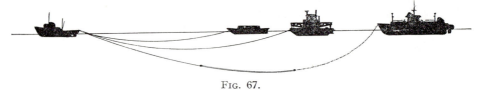

Fig. 67.

2. More than one Ocean-Going Tug Employed in the Towage of a Single Large Vessel or other Large Inanimate Floating Object.

From time to time very large vessels without power and other huge inanimate objects such as Floating Dry Docks, Oil Drilling Rigs, Pre-Fabricated Sea Defence Units, Sea Marks and the like require transportation by towage overseas and it is quite often the case that the bulk and form of such objects present something beyond the individual capacity of such tugs as may be available for the operation so that, if the unit is to be towed at a safe navigational speed and handled and manoeuvred to safe seamanlike standards, then the employment of more than one tug upon the operation becomes entirely mandatory. Second only to this requirement comes the necessity for applying towing power to the unit concerned in a balanced form.

Any lack of balance in the application of towing power in an operation soon becomes evident, regardless of the number of tugs engaged, because of difficulties in making good courses and speeds and in marked tendencies towards either yaw or sheer. This shows most clearly when two tugs of differing power are employed upon the towage of a single unit. Under such conditions the unit in tow tends to fall in astern of the tug which exerts the greater effort, this causes the towing medium from the weaker tug to apply obliquely towards the unit towed to the detriment of the mechanical efficiency of her effort, a condition which will clearly further deteriorate with the poor station keeping which this condition will engender. The only possible remedy in such cases is to reduce the effort of the stronger tug until a balance is reached, this a most uneconomic proceeding. Whilst the results of imbalanced effort illustrate more easily in the case of operations concerning two tugs, the evidence deriving from performance figures and passage behaviour of operations involving three and four tugs, where similar aggregates of power have been employed in varying combinations in the towage of similar units, shows repeated and conclusive proof of the advantages which accrue from the use of a balanced and symetrical towing force.

When the unit to be towed is a large vessel, the number of tugs to be employed will be restricted to the number of efficient towage points which are available in that vessel. Seeing that long experience of such operations has demonstrated the essentiality of effecting towage connections via hawse pipes and chain cables, it follows that the number of tugs to be employed will be two because the number of vessels, regardless of type and nationality, which can offer more than two hawse facilities simultaneously is now quite negligible. In providing a balanced towing effort two tugs of identical class would, of course, provide the ideal force but when this is not practicable tugs may be employed of differing classes providing that the discrepancy between their respective outputs does not exceed ten per cent.

A large unit of other than ship-shape in tow of two tugs usually utilises a pair of chain cable towing pennants of such length and dimensions, and each provided with a suitable ring, that they can together be formed up into a towing bridle for the use of either of the two tugs for convenience and flexibility whilst entering port and leaving, and to serve in an emergency should one of the pair become incapacitated.

Occasionally units present for overseas towage where, for operational or other related reasons, it is not possible to arrange towage points upon, or within, the fabric of the unit proper so that a towing bridle is required. When these units are of such size as to require two tugs for effective propulsion then very special conditions apply. If more than one tug is required to exert effort via a common bridle then their respective towing media must combine at the apex of the bridle utilising a ring of appropriate proportions and dimensions. In seamanlike consideration of this circumstance it is submitted that whilst the connecting shackles from two separate towing media might well be safely and efficiently accommodated into a ring of suitable dimensions, *more than two could not,* bearing in mind the fact that the ring will already be accommodating the shackles, of equal size and character, needed for the formation of the towage bridle, and also remembering the necessity of providing a fair lead and a safe bearing for each of the shackles in use under the practical conditions obtaining in day to day all weather towage.

The number of tugs to be employed upon towage operations involving very large other than ship-shaped objects, will invariably depend upon two factors:—

1. The number of satisfactory towing points which can be arranged upon the unit in question.

2. The number and class of tugs available.

Ideally, of course, the towing force would be made up of the appropriate number of tugs of identical class utilising identical towing media, and this sometimes is possible and when it does the operation is usually efficiently concluded to everyone's satisfaction for a great number of reasons other than those which are readily apparent. It is rare indeed to discover circumstances which demand the services of more than four tugs simultaneously and now that very large and very powerful units are available, most units offering for overseas towage fall within the competence of two or three tugs' combined capacities.

When, however, four tugs are employed upon an operation and when they cannot be of identical class, every effort is made to provide the necessary power with two pairs of tugs. The more powerful pair are assigned to the inner positions with the weaker units occupied at the wings. Each tug connects her towing gear into a chain cable, suitably made up to the tow, of identical lengths and the centre pair of pennants are arranged as was described above to form up into a bridle should this become necessary for any one of the reasons that have been offered.

An arrangement of three tugs has become exceedingly popular for long overseas towage operations where the required aggregate of power can be thus interpreted. Such a force comprises a pair of sisters and one odd class tug which may be either smaller or greater than the sister vessels but which is usually the latter. The odd vessel assumes the centre, or pole, position towing upon a bridle whilst the sister tugs take up the wing positions, utilising, as before, chain pennants of a convenient length. This arrangement is a thoroughly seamanlike one; the chain bridle and pennants add worthy weight to the towing media and absorb a great deal of unavoidable punishment inflicted by the terminal and staging manoeuvring, and which might not be so easily sustained by even the best steel wire ropes. The centre tug to a bridle permits of full control whilst forming up or shortening in, whilst it also allows this tug to be detached conveniently, should it become necessary for berthing, etc., leaving the unit in the competent charge of the wing tugs.

It is a fact, however, that the choice of tugs is sometimes far from the ideal, and it is also the truth that a great variety of circumstances may apply which enjoin the employment of tugs in such combinations as would not otherwise be acceptable. This happens when their is only one practicable point of tow upon a unit, and where no single tug available is competent to the operation so that the services of two must be utilised. Similar conditions apply when, in search of a balanced effort from an ill matched selection, two tugs are obliged to share a single medium.

The practice which such circumstances obliges, and which is known in the calling as 'Tandem Towage', consists of making a towage connection between one tug and the unit to be towed and then to augment the effort thus provided by connecting the sea towing gear of a second tug to the foreparts of the first. As with 'In Line' Towage, as earlier described, the excessive operational length which results from these practices is not navigationally desirable upon a number of counts. Besides these a deal of inefficiency and severe discomfort is inflicted upon the tug 'Working in Shafts', (as the middle position has been named,) because she must work in the screw race of the tug ahead of her and because of the inhibitions to her free movement either in the seaway, or in answer to her helm, because of being pinned at both ends. Further to this, this mode of working requires the tug 'In Shafts' to make use of towing gear which is competent to the combined effort of two tugs, gear of such dimensions as cannot possibly match her towing winch so that she will, almost inevitably, be deprived of this facility. It is also probable that the increased weight of gear may well be too much for her handling equipment to the detriment of her performance at the commencement, conclusion and staging points of the voyage.

Tugs which are involved in complex towage operations requiring two, three or four tugs must obviously adhere most rigidly and strictly to the routine practices and drill such as were outlined in Section IV, Chapter XII; but seeing that every action and reaction obtaining in such operations cannot be considered save in their effect upon the unit in tow and the other tug or tugs involved, such operations clearly require an overall Commander if maximum safety and full efficiency are to be enjoyed. It is, therefore, the normal procedure to appoint a Tug Commodore to assume overall command of these operations. This Officer will ordinarily be the Senior Tug Commanding Officer present and usually commands one of the tugs employed. When three tugs are employed he exercises his control from the centre position and if four tugs are present his tug is commonly one of the centre pair. Such a Commodore, being provided with the fullest details of the other tugs, both in performance and equipment, will order the route to be followed, the scope of towing media to be used, and the power to be employed; according to his interpretation of the physical conditions of wind and weather, sea and swell, as they obtain from day to day throughout the operation. He will be responsible for the navigation of the complex being assisted, in the matter of position finding, by the navigators borne in the other tugs and in the unit in tow.

The common practice upon such operations is for the Commodore to veer the scope of towing medium from his own tug which he considers to be adequate to the occasion thereafter ordering the other tugs in company to take station upon him. This is achieved by each tug veering gear until she comes up to a position a little abaft the Commodore's beam so that a constant visual check is allowed. Subsequent adjustments to the scope of gear employed are then signalled, by the means available, from the commodore tug to be implemented upon the transmission of an executive signal.

Because towage efficiency demands that each tow-rope in use shall lay as fairly parallel to the line of advance as may be practicable, close and precise station keeping is wholly obligatory upon each tug. This kind of long term close quarters navigation is somewhat unnerving at first, particularly under conditions of adverse weather, but when the involved personnel have had time to accustom themselves to the conditions and have learned that the risk of collision between the tugs is slight, and can, in effect, be allowed from time to time, then it becomes an accepted condition of employment and causes no further unease. (Fig. No. 68).

In actual practice, although the Commodore will promulgate his courses, and will signal all changes, his associated tugs will concentrate more upon the maintenance of proper station and devoting steering interest toward this rather than aiming towards the maintenance of individual compass courses. The growing utilisation of automatic steering devices has done a very great deal towards the solution of steering and station keeping problems so that it is now increasingly the custom, when these appliances are in use, for Commodores to signal their steering gear settings from time to time for the guidance of the others.

The Commodore's signals in regard to engine revolutions are however meticulously obeyed so as to preserve the balanced towage effort which is so essential to the success of operations.

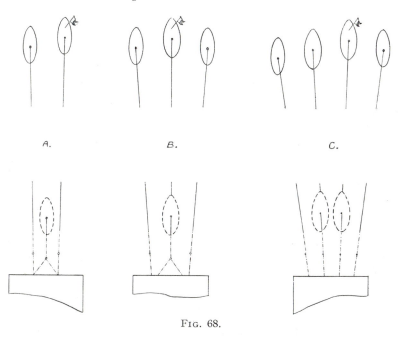

A.   B.   C.

Fig. 68.

Large changes of both speed and direction, such as may be imposed by conditions of traffic or emergency, are always executed under the personal command of the Commodore. When this sort of action is obliged the Commodore will signal his requirements, including the new course to be steered. When this order has been acknowledged by all, including the unit in tow, the executive signal is made. The Commodore himself will not, however, commence the manoeuvre, nor will he allow any movement from the tug or tugs lying on his side towards the direction of the new course to be assumed, but he will allow the initiative to be taken by his outside wing tug whereupon the other tugs will swing over in concert. This procedure is most essential because, particularly when very large units are in tow, the tendency is for the tugs to swing in arc upon the radius provided by their towing media, about a centre provided by the unit towed which will begin to lose way the moment the swing begins. Under these conditions the tug or tugs on the outer stations are required to make a larger movement to assume the new heading whilst they are also supporting more of the stress of towage. It will also be apparent, upon consideration, that these tugs will also have to work in the screw race of the others, once the proper aspects for the desired alterations have been assumed, and will be obliged to suffer this until the swing is completed.

Emergency movements such as this are clearly not desirable in view of the substantial increase in hazard which attends; for this reason a very sharp lookout has become a feature of all ocean tug routine. Emergencies must nevertheless attend upon all marine operations upon occasion, and in the case of major towage operations the overall efficiency of the unit must never be reduced by hesitation or haste, panic or indiscipline on the part of the Commanding Officers, Officers and ratings involved. The Commodore, for his part, must never order the execution of orders until he has an unqualified comprehension from his associated tugs. Under no circumstances should any tug deviate from the orders given, once they have been acknowledged, this because it is a supremely simple matter for any untoward movement, during one of these complex manoeuvres, to occasion disaster to the tug concerned or to any of her sisters, through collision, the fouling of a tug's own screw upon her own gear, the fouling of a sister's gear upon her own gear, or to foul the screws of other tugs in company. Further risk entailed is that of 'Girding', a danger which always attends in the wake of misadventure in towing.

The phenomenon known as 'Girding' has already been touched upon in other parts of this work but advantage must be taken of this chapter to more fully explain it. This condition is sometimes expressed as 'Girting' or 'Binding,' depending upon the origins of the tug or her crew. A tug is said to be 'Girded', 'Girted' or 'Bound' when her tow-rope leads out abeam, or before the beam, whilst enduring stress sufficient to entail a risk of capsizing, and when the action of the tug's helm and engines cannot serve to overcome the danger. Whilst this hazard is the common lot of harbour tugs, because of the complication of their manoeuvres, and because the vessels which they assist usually have main power which may be either accidentally or inadvertently used to produce a capsizing effect to any tug in attendance which may be holding an attitude broadside on to the ship's line of advance, this hazard does not unduly concern ocean-going tugs engaged in single unit towage because of the basic simplicity of their manoeuvres and because any way induced into the unit in tow will have been initiated by the tug herself and which should therefore ordinarily be controllable. When tugs engage in the more complex operations however, the risk of 'Girding' emerges for the reason that there are other forces in action upon any unit in tow other than those being imparted by any one individual tug.

Whilst the capsizing of tugs, under any set of circumstances, invariably occasions danger to life, limb and property, this accident to a tug upon the high seas almost invariably causes substantial loss of life, sometimes all hands, and the total loss of the tug herself. This because of the catastrophic nature of the accident and the rapid succession of events from initial causes to the shocking culmination, circumstances which are never improved by physical difficulties which result within the hull.

For this reason the highest possible degree of vigilance must be maintained in tugs engaged in complex operations in regard to all of the circumstances which are well known as being contributory to 'Girding'. Besides this every measure which may contribute toward the preservation of the tug and her ship's company in the face of this hazard must be the subject of both instruction and practice. Every Officer and Rating in the tug should, for instance, be aquainted with the means available for effecting the rapid release of the towing medium whether towing proceeds by means of a towing winch, towing hook or towing bollard. The content of Section IV, Chapter XI is particularly applicable in this context.

R

APPENDICES

STANDARD FORM OF

SALVAGE AGREEMENT

(APPROVED AND PUBLISHED BY THE COMMITTEE OF LLOYD'S)

NO CURE——NO PAY.

On board the

Dated   19

† See Note 1
above

* See Note 2
above

** See Note 3
above

IT IS HEREBY AGREED between Captain†   for and on
behalf of the Owners of the "   " her Cargo and
Freight and   for and on behalf of
(hereinafter called "the Contractor"*) :—

1.   The Contractor agrees to use his best endeavours to salve the
and her cargo and take them into   or other place to be hereafter agreed
with the Master, providing at his own risk all proper steam and other assistance and labour.   The
services shall be rendered and accepted as salvage services upon the principle of "no cure—
no pay" and the Contractors remuneration in the event of success shall be **£   ,
unless this sum shall afterwards be objected to as hereinafter mentioned in which case the
remuneration for the services rendered shall be fixed by Arbitration in London in the manner
hereinafter prescribed and any other difference arising out of this Agreement or the operations
thereunder shall be referred to Arbitration in the same way.   In the event of the services referred
to in this Agreement or any part of such services having been already rendered at the date of
this Agreement by the Contractor to the said vessel or her cargo it is agreed that the provisions
of this Agreement shall *mutatis mutandis* apply to such services.

2.   The Contractor may make reasonable use of the vessel's gear anchors chains and other
appurtenances during and for the purpose of the operations free of costs but shall not unnecessarily
damage abandon or sacrifice the same or any other of the property.

3. Notwithstanding anything hereinbefore contained should the operations be only partially successful without any negligence or want of ordinary skill and care on the part of the Contractor or of any person by him employed in the operations, and any portion of the Vessel's Cargo or Stores be salved by the Contractor, he shall be entitled to reasonable remuneration not exceeding a sum equal to   per cent of the estimated value of the property salved at or if the property salved shall be sold there then not exceeding the like percentage of the net proceeds of such sale after deducting all expenses and customs duties or other imposts paid or incurred thereon but he shall not be entitled to any further remuneration reimbursement or compensation whatsoever and such reasonable remuneration shall be fixed in case of difference by Arbitration in manner hereinafter prescribed.

4. The Contractor shall immediately after the termination of the services or sooner notify the Committee of Lloyd's of the amount for which he requires security to be given; and failing any such notification by him not later than 48 hours (exclusive of Sundays or other days observed as general holidays at Lloyd's) after the termination of the services he shall be deemed to require security to be given for the sum named in Clause 1, or, if no sum be named in Clause 1, then for such sum as the Committee of Lloyd's in their absolute discretion shall consider sufficient. Such security shall be given in such manner and form as the Committee of Lloyd's in their absolute discretion may consider sufficient but the Committee of Lloyd's shall not be in any way responsible for the sufficiency (whether in amount or otherwise) of any security accepted by them nor for the default or insolvency of any person firm or corporation giving the same.

5. Pending the completion of the security as aforesaid, the Contractor shall have a maritime lien on the property salved for his remuneration. The salved property shall not without the consent in writing of the Contractor be removed from   or the place of safety to which the property is taken by the Contractor on the completion of the salvage services until security has been given to the Committee of Lloyd's as aforesaid. The Contractor agrees not to arrest or detain the property salved unless the security be not given within 14 days (exclusive of Sundays or other days observed as general holidays at Lloyd's) of the termination of the services (the Committee of Lloyd's not being responsible for the failure of the parties concerned to provide the required security within the said 14 days) or the Contractor has reason to believe that the removal of the property salved is contemplated contrary to the above agreement. In the event of security not being provided as aforesaid or in the event of any attempt being made to remove the property salved contrary to this agreement the Contractor may take steps to enforce his aforesaid lien. The Arbitrator or Arbitrators or Umpire (including the Committee of Lloyd's if they act in either capacity) appointed under Clauses 7 or 8 hereof shall have power in their absolute discretion to include in the amount awarded to the Contractor the whole or such part of the expenses incurred by the Contractor in enforcing his lien as they shall think fit.

6. After the expiry of 42 days from the date of the completion of the security the Committee of Lloyd's shall call upon the party or parties concerned to pay the amount thereof and in the event of non-payment shall realize or enforce the security and pay over the amount thereof to the Contractor unless they shall meanwhile have received written notice of objection and a claim for Arbitration from any of the parties entitled and authorized to make such objection and claim or unless they shall themselves think fit to object and demand Arbitration. The receipt of the Contractor shall be a good discharge to the Committee of Lloyd's for any monies so paid and they shall incur no responsibility to any of the parties concerned by making such payment and no objection or claim for Arbitration shall be entertained or acted upon unless received by the Committee of Lloyd's within the 42 days above mentioned.

7. In case of objection being made and Arbitration demanded the remuneration for the services shall be fixed by the Committee of Lloyd's as Arbitrators or at their option by an Arbitrator to be appointed by them unless they shall within 30 days from the date of this Agreement receive from the Contractor a written or telegraphic notice appointing an Arbitrator

1-08
2-24
0-26
4-50
6-53

Z11

on his own behalf in which case such notice shall be communicated by them to the Owners of the vessel and they shall within 15 days from the receipt thereof give a written notice to the Committee of Lloyd's appointing an Arbitrator on behalf of all the parties interested in the property salved; and if the Owners shall fail to appoint an Arbitrator as aforesaid the Committee of Lloyd's shall appoint another Arbitrator on behalf of all the parties interested in the property salved or they may if they think fit direct that the Contractor's nominee shall act as sole Arbitrator; and thereupon the Arbitration shall be held in London by the Arbitrators or Arbitrator so appointed. If the Arbitrators cannot agree they shall forthwith notify the Committee of Lloyd's who shall thereupon either themselves act as Umpires or shall appoint some other person as Umpire. Any award of the Arbitrators or Arbitrator or Umpire shall (subject to appeal as provided in this Agreement) be final and binding on all the parties concerned and they or he shall have power to obtain call for receive and act upon any such oral or documentary evidence or information (whether the same be strictly admissible as evidence or not) as they or he may think fit, and to conduct the Arbitration in such manner in all respects as they or he may think fit, and to maintain reduce or increase the sum, if any, named in Clause 1, and shall if in their or his opinion the amount of the security demanded is excessive have power in their or his absolute discretion to condemn the Contractor in the whole or part of the expense of providing such security and to deduct the amount in which the Contractor is so condemned from the salvage remuneration. Unless the Arbitrators or Arbitrator or Umpire shall otherwise direct the parties shall be at liberty to adduce expert evidence on the Arbitration. The Arbitrators or Arbitrator and the Umpire (including the Committee of Lloyd's if they act in either capacity) may charge such fees as they may think reasonable, and the Committee of Lloyd's may in any event charge a reasonable fee for their services in connection with the Arbitration, and all such fees shall be treated as part of the costs of the Arbitration and Award and shall be paid by such of the parties as the Award may direct. Interest at the rate of 5 per cent per annum from the expiration of 14 days (exclusive of Sundays or other days observed as general holidays at Lloyd's) after the date of the publication of the Award by the Committee of Lloyd's until the date of payment to the Committee of Lloyd's shall (subject to appeal as provided in this Agreement) be payable to the Contractor upon the amount of any sum awarded after deduction of any sums paid on account. Save as aforesaid the statutory provisions as to Arbitration for the time being in force in England shall apply. The said Arbitration is hereinafter in this Agreement referred to as "the original Arbitration" and the Arbitrator or Arbitrators or Umpire thereat as "the original Arbitrator" or "the original Arbitrators" or "the Umpire" and the Award of such Arbitrator or Arbitrators or Umpire as "the original Award".

8. Any of the persons named under Clause 14, except the Committee of Lloyd's may appeal from the original Award by giving written Notice of Appeal to the Committee of Lloyd's within 14 days (exclusive of Sundays or other days observed as general holidays at Lloyd's) from the publication by the Committee of Lloyd's of the original Award; and any of the other persons named under Clause 14, except the Committee of Lloyd's, may (without prejudice to their right of appeal under the first part of this clause) within 7 days (exclusive of Sundays or other days observed as general holidays at Lloyd's) after receipt by them from the Committee of Lloyd's of notice of such appeal (such notice if sent by post to be deemed to be received on the day following that on which the said notice was posted) give written Notice of Cross-Appeal to the Committee of Lloyd's. As soon as practicable after receipt of such notice or notices the Committee of Lloyd's shall themselves alone or jointly with another person or other persons appointed by them (unless they be the objectors) hear and determine the Appeal or if they shall see fit to do so or if they be the objectors they shall refer the Appeal to the hearing and determination of a person or persons selected by them. Any award on Appeal shall be final and binding on all the parties concerned. No evidence other than the documents put in on the original Arbitration and the original Arbitrator's or original Arbitrators' and/or Umpire's notes and/or shorthand notes if any of the proceedings and oral evidence if any at the original Arbitration shall be used on the Appeal unless the Arbitrator or Arbitrators on the Appeal shall in his or their discretion call for other evidence. The Arbitrator or Arbitrators on the Appeal may conduct the Arbitration on Appeal

in such manner in all respects as he or they may think fit and may maintain increase or reduce the sum awarded by the original Award with the like power as is conferred by Clause 7 on the original Arbitrator or Arbitrators or Umpire to condemn the Contractor in the whole or part of the expense of providing security and to deduct the amount disallowed from the salvage remuneration. And he or they shall also make such order as he or they may think fit as to the payment of interest (at the rate of 5 per cent per annum) on the sum awarded to the Contractor. The Arbitrator or Arbitrators on Appeal (including the Committee of Lloyd's if they act in that capacity) may direct in what manner the costs of the original Arbitration and of the Arbitration on Appeal shall be borne and paid and may charge such fees as he or they may think reasonable and the Committee of Lloyd's may in any event charge a reasonable fee for their services in connection with the Arbitration on Appeal and all such fees shall be treated as part of the costs of the Arbitration and Award on Appeal and shall be paid by such of the parties as the Award on Appeal shall direct. Save as aforesaid the statutory provisions as to Arbitration for the time being in force in England shall apply.

9. (a) In case of Arbitration if no notice of Appeal be received by the Committee of Lloyd's within 14 days after the publication by the Committee of the original Award the Committee shall call upon the party or parties concerned to pay the amount awarded and in the event of non-payment shall realize or enforce the security and pay therefrom to the Contractor (whose receipt shall be a good discharge to them) the amount awarded to him together with interest as hereinbefore provided.

(b) If notice of Appeal be received by the Committee of Lloyd's in accordance with the provisions of Clause 8 hereof they shall as soon as but not until the Award on Appeal has been published by them, call upon the party or parties concerned to pay the amount awarded and in the event of non-payment shall realize or enforce the security and pay therefrom to the Contractor (whose receipt shall be a good discharge to them) the amount awarded to him together with interest if any in such manner as shall comply with the provisions of the Award on Appeal.

(c) If the Award on Appeal provides that the costs of the original Arbitration or of the Arbitration on Appeal or any part of such costs shall be borne by the Contractor, such costs may be deducted from the amount awarded before payment is made to the Contractor by the Committee of Lloyd's, unless satisfactory security is provided by the Contractor for the payment of such costs.

(d) Without prejudice to the provisions of Clause 4 hereof, the Liability of the Committee of Lloyd's shall be limited in any event to the amount of security held by them.

10. The Committee of Lloyd's may in their discretion out of the security (which they may realize or enforce for that purpose) pay to the Contractor on account before the publication of the original Award and/or of the Award on Appeal such sum as they may think reasonable on account of any out-of-pocket expenses incurred by him in connection with the services.

11. The Master or other person signing this Agreement on behalf of the property to be salved is not authorized to make or give and the Contractor shall not demand or take any payment draft or order for or on account of the remuneration.

12. Any dispute between any of the parties interested in the property salved as to the proportions in which they are to provide the security or contribute to the sum awarded or as to any other such matter shall be referred to and determined by the Committee of Lloyd's or by some other person or persons appointed by the Committee whose decision shall be final and is to be complied with forthwith.

13. The Master or other person signing this Agreement on behalf of the property to be salved enters into this Agreement as Agent for the vessel her cargo and freight and the respective owners thereof and binds each (but not the one for the other or himself personally) to the due performance thereof.

14. Any of the following parties may object to the sum named in Clause 1 as excessive or insufficient having regard to the services which proved to be necessary in performing the Agreement or to the value of the property salved at the completion of the operations and may claim Arbitration viz :—(1) The Owners of the ship. (2) Such other persons together interested as Owners and/or Underwriters of any part not being less than one-fourth of the estimated value of the property salved as the Committee of Lloyd's in their absolute discretion may by reason of the substantial character of their interest or otherwise authorize to object. (3) The Contractor. (4) The Committee of Lloyd's—Any such objection and the original Award upon the Arbitration following thereon shall (subject to appeal as provided in this Agreement) be binding not only upon the objectors but upon all concerned, provided always that the Arbitrators or Arbitrator or Umpire may in case of objection by some only of the parties interested order the costs to be paid by the objectors only, provided also that if the Committee of Lloyd's be objectors they shall not themselves act as Arbitrators or Umpires.

15. If the parties to any such Arbitration or any of them desire to be heard or to adduce evidence at the original Arbitration they shall give notice to that effect to the Committee of Lloyd's and shall respectively nominate a person in the United Kingdom to represent them for all the purposes of the Arbitration and failing such notice and nomination being given the Arbitrators or Arbitrator or Umpire may proceed as if the parties failing to give the same had renounced their right to be heard or adduce evidence.

16. Any Award, notice, authority, order, or other document signed by the Chairman of Lloyd's or a Clerk to the Committee of Lloyd's on behalf of the Committee of Lloyd's shall be deemed to have been duly made or given by the Committee of Lloyd's and shall have the same force and effect in all respects as if it had been signed by every member of the Committee of Lloyd's.

For and on behalf of the Contractor

. . . .

(To be signed either by the Contractor personally or by the Master of the salving vessel or other person whose name is inserted in line 3 of this Agreement.)

For and on behalf of the Owners of property to be salved

. . . .

(To be signed by the Master or other person whose name is inserted in line 1 of this Agreement.)

The Open Form was first published in the year 1892, but has, since then, been amended several times. Possibly the most significant amendment was the introduction, in the year 1926, of a clause providing for an Appeal in the case that one, or all, of the interested parties was dissatisfied with an award arising out of arbitration arising out of an implementation of the Agreement. Prior to 1926 an Arbitrator's judgement was wholly binding upon all of the interested parties.

Although it is generally held to be desirable that both Owners and Salvors should sign a Form of Agreement before salvage services commence they may also sign, if they so choose, when the services are completed.

Whilst it is obviously preferable that the concerned parties should sign the proper form of agreement, ample precedent exists for the acceptance of a simple written agreement on the part of the interested parties that both accept the terms of Lloyd's Open Form. It has, moreover, been offered that an exchange of radio messages expressing agreement with the terms of the Form is equally effective, always provided that they are correctly presented upon the appropriate Telegram Forms and are properly logged.

Whilst the Open Form provides space for the inclusion of the amount of financial remuneration, it is the ordinary practice not to insert any figure in view of the fact that, in practice, the ultimate award will be determined in arbitration.

The Agreement does, however, allow that the Salvor should notify the amount of security required and that he may retain his maritime lien upon the ship and property salved, which may not be moved without his consent. On the other hand the Agreement obliges the Salvor to agree not to arrest the salved property if the required security is provided within fourteen days.

If, as is customary, arbitration is demanded by any or all of the interested parties, the Committee of Lloyd's will, upon request, serve in this capacity themselves or appoint a proper person to the task. The interested parties may, of course, appoint their own choice.

Now that Appeal is allowed, the Committee of Lloyd's will themselves serve in this capacity or appoint proper persons to the task. Arbitrators selected and appointed by the Committee for this purpose are Lawyers experienced in Admiralty and Maritime Law generally and procedures are governed by the general law of England. Arbitrators are obliged to apply the English Law of Salvage so that the interested parties are assured of proper judgements with awards to recognised standards.

APPENDIX II

Table 1

Detail of the Construction of Cable Laid Fibre Towing Ropes

Circumference of the rope in inches	Diameter of the rope in mms.	Number of yarns in each Primary Strand
10″	80 mm.	63
12″	96 mm.	91
14″	112 mm.	124
16″	128 mm.	163
18″	144 mm.	205
20″	160 mm.	253
22″	176 mm.	306
24″	192 mm.	364
26″	208 mm.	428

Table 2

Detail of the Construction of Steel Wire Towing Ropes

A. **Ropes having six strands each with thirty-seven wires** (6×37) .... $(18/12/6)1$

Number of Wires per Rope			Reserve Strength Inner Wires	Remaining Strength Outer Wires
Inners	Outers	Total		
114	108	222	51%	49%

B. Ropes having six strands each with sixty-one wires
(6 × 61) . . . . (24/18/12/6/1)

Number of Wires per Rope			Reserve Strength Inner Wires	Remaining Strength Outer Wires
Inners	Outers	Total		
222	144	366	60%	40%

Table 3

Detail of the Construction of Hawser Laid Nylon Towing Ropes

Circumference of the rope in inches	No. of Nylon Filaments per strand (1,000)	No. of Terylene Filaments per strand (1,000)
8″	853	1,308
9″	1080	1,656
10″	1330	2,044.5
11″	1613	2,474
12″	1920	2,944
13″	2253	3,455
14″	2613	4,007
15″	3000	4,600

A table giving the respective sizes of towing media which may be combined to provide a reasonably balanced towing set. It will be noted that the sizes of Steel Wire Rope and Studded Link Cable give breaking strains somewhat in excess of the soft media. This is a common towage practice which endeavours to make practical compensation for their incapacity for elasticity or extension.

Table 4

Manila Cable Laid Grade 1	Towing Quality Steel Wire Rope	Nylon Hawser Laid	Studded Link Anchor Cable
10″	3″ —3½″	6″	1⅛″
12″	3½″—4″	7″	1⅜″
14″	4″ —4½″	8″	1½″
16″	4½″—5″	9″	1¾″
18″	5″ —5½″	10″	1⅞″
20″	5½″—6″	11″	2″
22″	6″ —6½″	12″	2¼″
24″	6½″—7″	13″	2½″
26″	7″ —7½″	14″	2⅜″

Table 5

Approximate Weights, per fathom and per metre, of the Principal Ocean Towing Media

A. Cable Laid Grade 1 Manila

Circ. in Inches	Weight	
	Lbs.	Kilos
10″	15.5	3.4
12″	22.4	5.5
14″	30.6	7.6
16″	40.0	9.8
18″	50.6	12.5
20″	62.0	15.6
22″	77.5	19.5
24″	89.9	22.3
26″	102.4	25.5

B. Towage Quality Steel Wire Rope

Circumference in Inches	Weight	
	Lbs.	Kilos
3″	9.5	2.3
3½″	12.0	3.0
4″	14.9	3.7
4½″	18.2	4.5
5″	23.8	5.8
5½″	29.4	7.2
6″	37.8	9.5
6½″	42.0	10.8
7″	48.0	12.0
7½″	55.7	14.3

C. Three Stranded, Hawser Laid Nylon Towing Rope

Circumference in Inches	Weight	
	Lbs.	Kilos
6″	6.5	1.6
7″	8.8	2.2
8″	11.5	2.9
9″	14.5	3.6
10″	18.0	4.5
11″	21.6	5.3
12″	25.8	6.4
13″	30.2	7.4
14″	35.9	8.5
15″	41.8	9.6

D. Studded Link Anchor Cable
$(W = 55d^2)$

Diameter in Inches	Weight	
	Lbs.	Kilos
$1\frac{1}{8}''$	63.7	15.7
$1\frac{3}{8}''$	104.5	25.8
$1\frac{1}{2}''$	123.3	30.4
$1\frac{3}{4}''$	168.3	41.7
$1\frac{7}{8}''$	192.5	47.6
$2''$	220.0	54.4
$2\frac{1}{4}''$	280.5	69.5
$2\frac{3}{8}''$	308.0	76.2
$2\frac{1}{2}''$	343.7	85.1

Rule 5 (*b*)

Between sunrise and sunset a vessel being towed, if the length of the tow exceeds 600 feet shall carry where it can best be seen a black diamond shape at least 2 feet in diameter.

Rule 15 (*c*)

(v) A vessel when towing, a vessel engaged in laying or in picking up a submarine cable or navigation mark, and a vessel under way which is unable to get out of the way of an approaching vessel through being not under command or unable to manoeuvre as required by these Rules shall, instead of the signals prescribed in sub-sections (i), (ii) and (iii) sound, at intervals of not more than 1 minute, three blasts in succession, namely, one prolonged blast followed by two short blasts.

(vi) A vessel towed, or, if more than one vessel is towed, only the last vessel of the tow, if manned, shall, at intervals of not more than 1 minute, sound four blasts in succession, namely, one prolonged blast followed by three short blasts. When practicable, this signal shall be made immediately after the signal made by the towing vessel.

APPENDIX III

Extracts from the International Regulations for preventing Collisions at Sea.

Rule 3 (a)

A power-driven vessel when towing or pushing another vessel or seaplane shall, in addition to her sidelights, carry two white lights in a vertical line one over the other, not less than 6 feet apart, and when towing and the length of the tow, measuring from the stern of the towing vessel to the stern of the last vessel towed, exceeds 600 feet, shall carry three white lights in a vertical line one over the other, so that the upper and lower lights shall be the same distance from, and not less than 6 feet above or below, the middle light. Each of these lights shall be of the same construction and character and one of them shall be carried in the same position as the white light prescribed in Rule 2 (a) (i). None of these lights shall be carried at a height of less than 14 feet above the hull. In a vessel with a single mast, such lights may be carried on the mast.

Rule 3 (b)

The towing vessel shall also show either the stern light prescribed in Rule 10 or in lieu of that light a small white light abaft the funnel or aftermast for the tow to steer by, but such light shall not be visible forward of the beam.

Rule 3 (c)

Between sunrise and sunset a power-driven vessel engaged in towing, if the length of tow exceeds 600 feet, shall carry, where it can best be seen, a black diamond shape at least 2 feet in diameter.

Rule 4 (a)

A vessel which is not under command shall carry, where they can best be seen, and, if a power-driven vessel, in lieu of the lights prescribed in Rule 2 (a) (i) and (ii), two red lights in a vertical line one over the other not less than 6 feet apart, and of such a character as to be visible all round the horizon at a distance of at least 2 miles. By day, she shall carry in a vertical line one over the other not less than 6 feet apart, where they can best be seen, two black balls or shapes each not less than 2 feet in diameter.

Rule 5 (a)

A sailing vessel under way and any vessel or seaplane being towed shall carry the same lights as are prescribed in Rule 2 for a power-driven vessel or a seaplane under way, respectively, with the exception of the white lights prescribed therein, which they shall never carry. They shall also carry stern lights as prescribed in Rule 10, provided that vessels towed, except the last vessel of a tow, may carry, in lieu of such stern light, a small white light as prescribed in Rule 3 (b).